student study
ART NOTEBOOK

HUMAN ANATOMY & PHYSIOLOGY
sixth edition

John W. Hole, Jr.

WCB **Wm. C. Brown Publishers**
Dubuque, Iowa · Melbourne, Australia · Oxford, England

Book Team

Editor *Colin H. Wheatley*
Developmental Editor *Kristine M. Noel*
Production Editor *Catherine S. DiPasquale*
Designer *Anna Manhart*
Art Editor *Joseph P. O'Connell*
Photo Editor *Lori Hancock*
Permissions Coordinator *Gail I. Wheatley*

Wm. C. Brown Publishers
A Division of Wm. C. Brown Communications, Inc.

Vice President and General Manager *Beverly Kolz*
Vice President, Publisher *Kevin Kane*
Vice President, Director of Sales and Marketing *Virginia S. Moffat*
National Sales Manager *Douglas J. DiNardo*
Marketing Manager *Craig S. Marty*
Advertising Manager *Janelle Keeffer*
Director of Production *Colleen A. Yonda*
Publishing Services Manager *Karen J. Slaght*
Permissions/Records Manager *Connie Allendorf*

Wm. C. Brown Communications, Inc.

President and Chief Executive Officer *G. Franklin Lewis*
Corporate Senior Vice President, President of WCB Manufacturing *Roger Meyer*
Corporate Senior Vice President and Chief Financial Officer *Robert Chesterman*

Cover © Scott Chaney

The credits section for this book begins on page 133 and is considered an extension of the copyright page.

A Times Mirror Company

ISBN 0-697-22787-1

Printed in the United States of America by Wm. C. Brown Communications, Inc., 2460 Kerper Boulevard, Dubuque, IA 52001

10 9 8 7 6 5 4 3 2 1

TO THE STUDENT

The *Student Study Art Notebook* is designed to help you in your study of human anatomy and physiology. The notebook contains art taken directly from the text and overhead transparencies; thus you can take notes during lectures, or jot down comments as you are reading through the chapters.

The notebook is perforated and 3-hole punched so, if you wish, you can remove sheets and put them in a binder with other study or lecture notes. Any blank pages at the end of this notebook can be used for additional notes or drawings.

We hope this notebook, used along with your text, helps to make the study of the human body easier for you.

DIRECTORY OF NOTEBOOK FIGURES

TO ACCOMPANY HOLE,

HUMAN ANATOMY & PHYSIOLOGY, 6/e

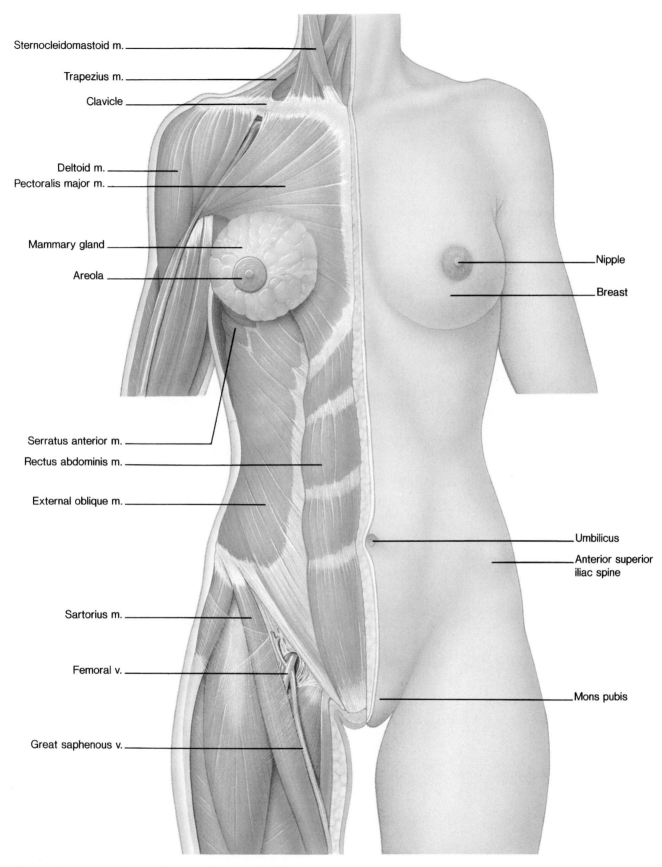

Sternocleidomastoid m.

Trapezius m.

Clavicle

Deltoid m.

Pectoralis major m.

Mammary gland

Areola

Serratus anterior m.

Rectus abdominis m.

External oblique m.

Sartorius m.

Femoral v.

Great saphenous v.

Nipple

Breast

Umbilicus

Anterior superior
iliac spine

Mons pubis

Human Torso, Anterior Surface and Superficial Muscles
Plate 1

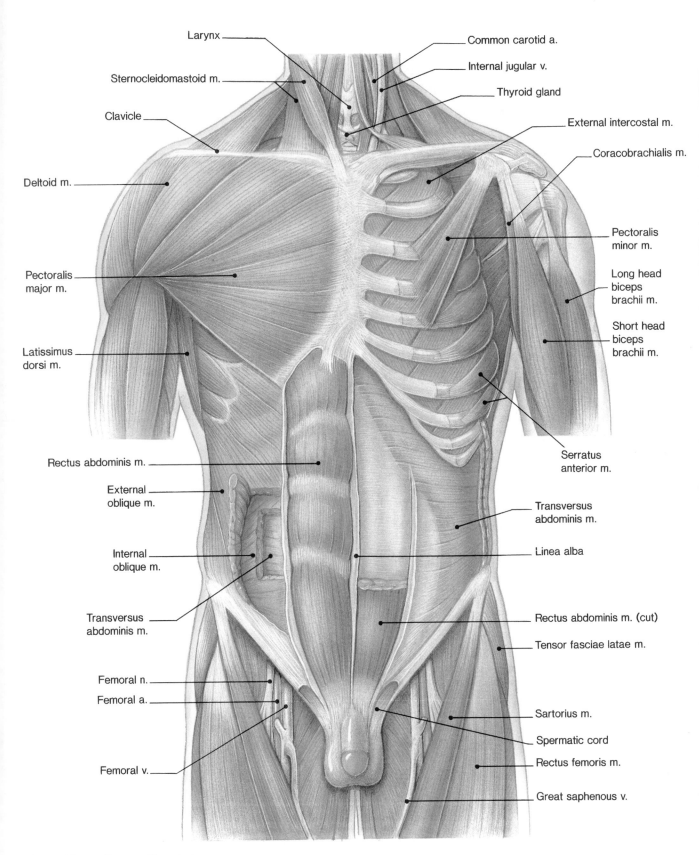

Larynx

Common carotid a.

Internal jugular v.

Sternocleidomastoid m.

Thyroid gland

Clavicle

External intercostal m.

Coracobrachialis m.

Deltoid m.

Pectoralis minor m.

Pectoralis major m.

Long head biceps brachii m.

Short head biceps brachii m.

Latissimus dorsi m.

Serratus anterior m.

Rectus abdominis m.

External oblique m.

Transversus abdominis m.

Linea alba

Internal oblique m.

Transversus abdominis m.

Rectus abdominis m. (cut)

Tensor fasciae latae m.

Femoral n.

Femoral a.

Sartorius m.

Spermatic cord

Rectus femoris m.

Femoral v.

Great saphenous v.

Human Torso, Deeper Muscle Layers
Plate 2

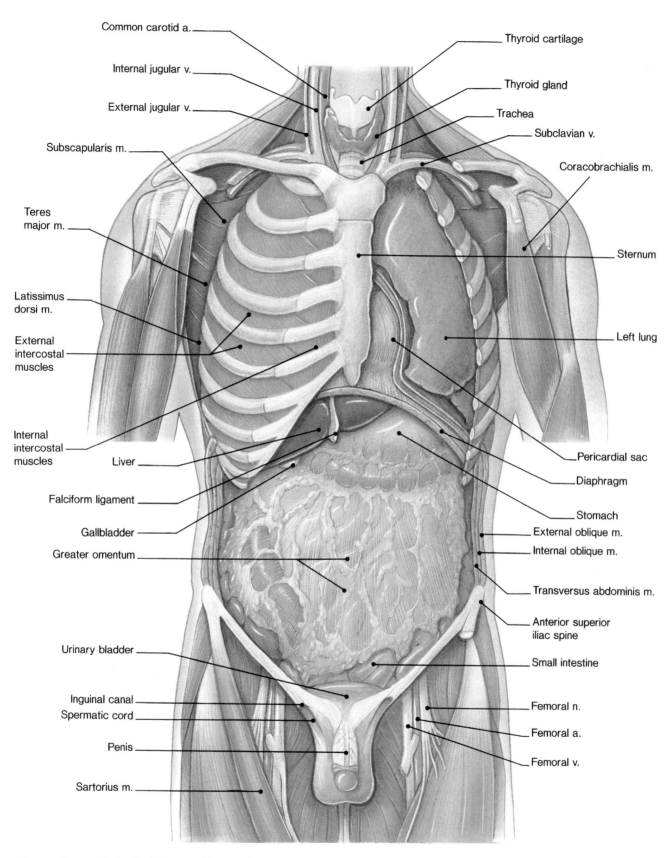

Common carotid a.

Internal jugular v.

External jugular v.

Subscapularis m.

Teres major m.

Latissimus dorsi m.

External intercostal muscles

Internal intercostal muscles

Liver

Falciform ligament

Gallbladder

Greater omentum

Urinary bladder

Inguinal canal

Spermatic cord

Penis

Sartorius m.

Thyroid cartilage

Thyroid gland

Trachea

Subclavian v.

Coracobrachialis m.

Sternum

Left lung

Pericardial sac

Diaphragm

Stomach

External oblique m.

Internal oblique m.

Transversus abdominis m.

Anterior superior iliac spine

Small intestine

Femoral n.

Femoral a.

Femoral v.

Human Torso, Abdominal Viscera Exposed
Plate 3

3

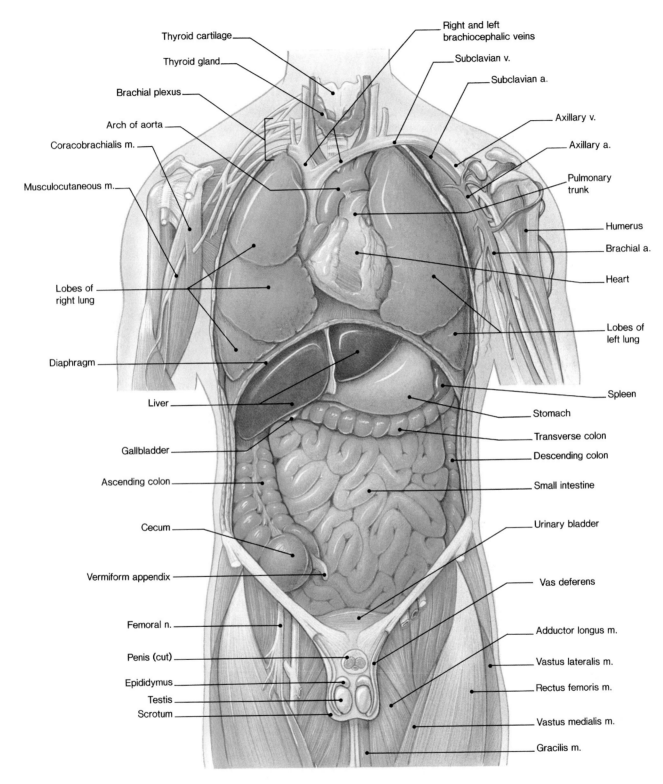

Thyroid cartilage

Thyroid gland

Brachial plexus

Arch of aorta

Coracobrachialis m.

Musculocutaneous m.

Lobes of right lung

Diaphragm

Liver

Gallbladder

Ascending colon

Cecum

Vermiform appendix

Femoral n.

Penis (cut)

Epididymus

Testis

Scrotum

Right and left brachiocephalic veins

Subclavian v.

Subclavian a.

Axillary v.

Axillary a.

Pulmonary trunk

Humerus

Brachial a.

Heart

Lobes of left lung

Spleen

Stomach

Transverse colon

Descending colon

Small intestine

Urinary bladder

Vas deferens

Adductor longus m.

Vastus lateralis m.

Rectus femoris m.

Vastus medialis m.

Gracilis m.

Human Torso, Thoracic Viscera Exposed
Plate 4

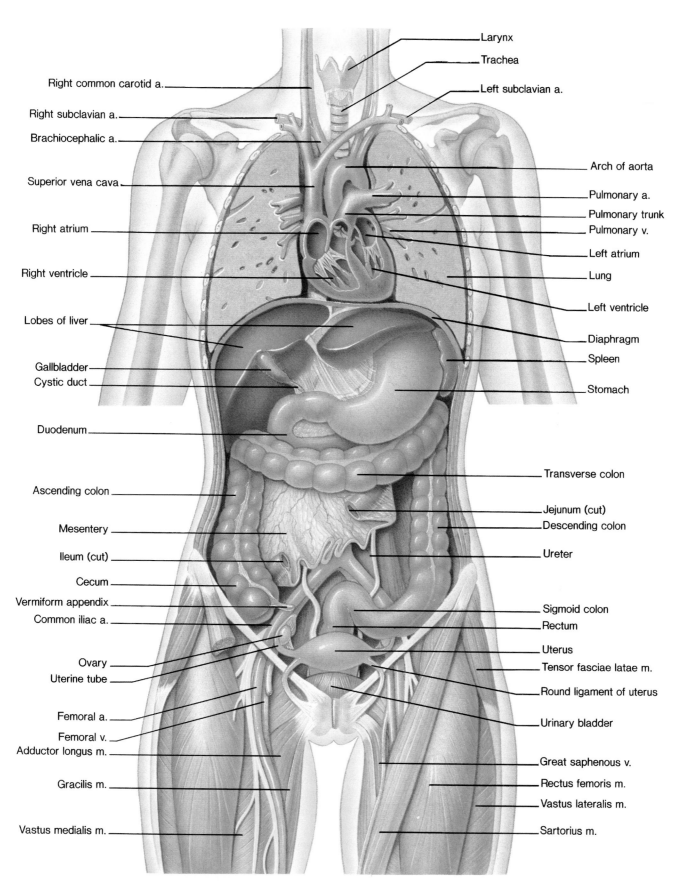

Right common carotid a.

Right subclavian a.

Brachiocephalic a.

Superior vena cava

Right atrium

Right ventricle

Lobes of liver

Gallbladder

Cystic duct

Duodenum

Ascending colon

Mesentery

Ileum (cut)

Cecum

Vermiform appendix

Common iliac a.

Ovary

Uterine tube

Femoral a.

Femoral v.

Adductor longus m.

Gracilis m.

Vastus medialis m.

Larynx

Trachea

Left subclavian a.

Arch of aorta

Pulmonary a.

Pulmonary trunk

Pulmonary v.

Left atrium

Lung

Left ventricle

Diaphragm

Spleen

Stomach

Transverse colon

Jejunum (cut)

Descending colon

Ureter

Sigmoid colon

Rectum

Uterus

Tensor fasciae latae m.

Round ligament of uterus

Urinary bladder

Great saphenous v.

Rectus femoris m.

Vastus lateralis m.

Sartorius m.

Human Torso, Lungs, Heart, and Small Intestine
Sectioned
Plate 5

5

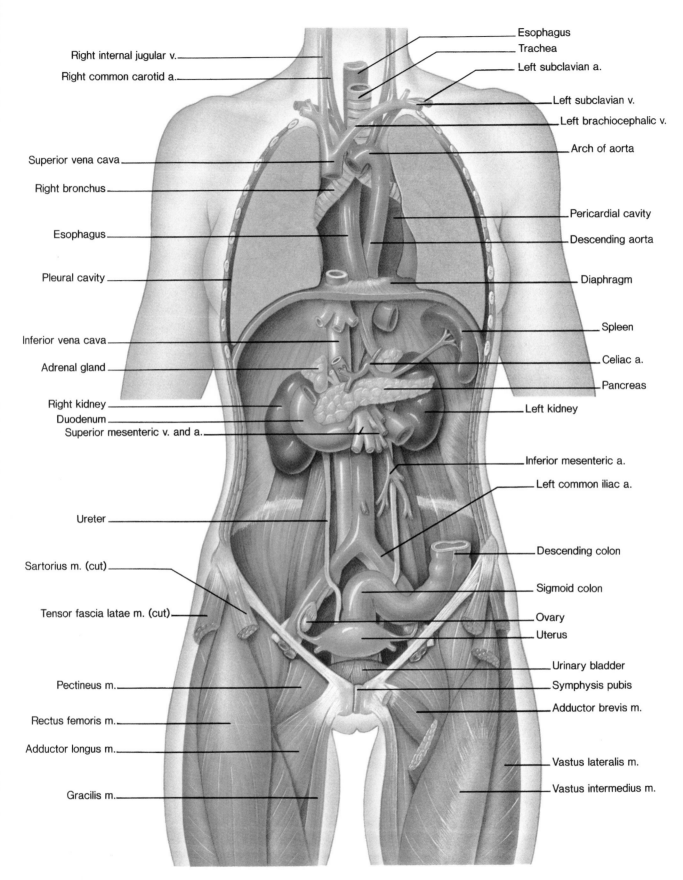

Right internal jugular v.

Right common carotid a.

Superior vena cava

Right bronchus

Esophagus

Pleural cavity

Inferior vena cava

Adrenal gland

Right kidney

Duodenum

Superior mesenteric v. and a.

Ureter

Sartorius m. (cut)

Tensor fascia latae m. (cut)

Pectineus m.

Rectus femoris m.

Adductor longus m.

Gracilis m.

Esophagus

Trachea

Left subclavian a.

Left subclavian v.

Left brachiocephalic v.

Arch of aorta

Pericardial cavity

Descending aorta

Diaphragm

Spleen

Celiac a.

Pancreas

Left kidney

Inferior mesenteric a.

Left common iliac a.

Descending colon

Sigmoid colon

Ovary

Uterus

Urinary bladder

Symphysis pubis

Adductor brevis m.

Vastus lateralis m.

Vastus intermedius m.

Human Torso, Heart, Stomach, and Parts of Intestine
and Lungs Removed
Plate 6

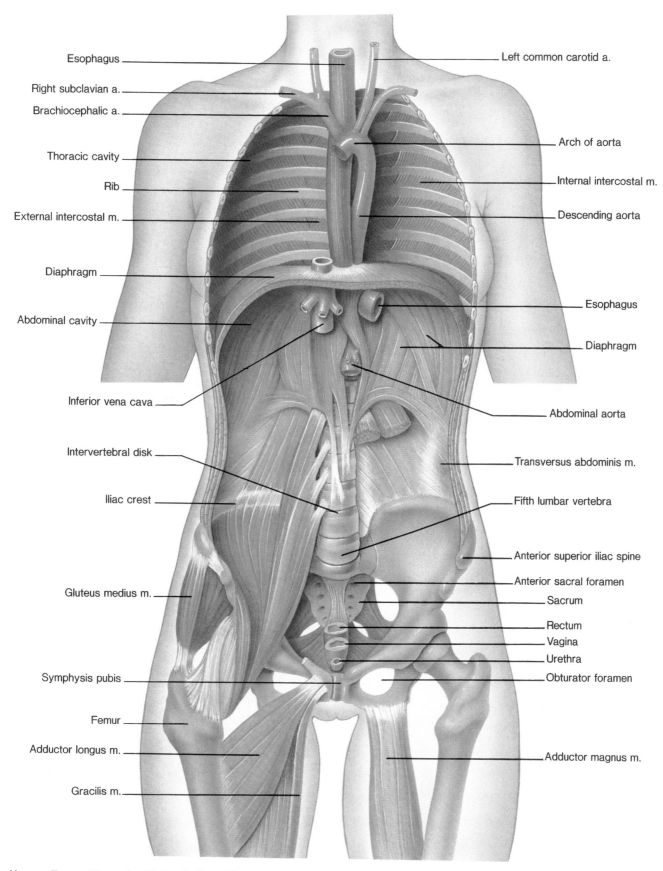

Esophagus —

Right subclavian a. —

Brachiocephalic a. —

Thoracic cavity —

Rib —

External intercostal m. —

Diaphragm —

Abdominal cavity —

Inferior vena cava —

Intervertebral disk —

Iliac crest —

Gluteus medius m. —

Symphysis pubis —

Femur —

Adductor longus m. —

Gracilis m. —

— Left common carotid a.

— Arch of aorta

— Internal intercostal m.

— Descending aorta

— Esophagus

— Diaphragm

— Abdominal aorta

— Transversus abdominis m.

— Fifth lumbar vertebra

— Anterior superior iliac spine

— Anterior sacral foramen

— Sacrum

— Rectum

— Vagina

— Urethra

— Obturator foramen

— Adductor magnus m.

Human Torso, Thoracic, Abdominal and Pelvic Visceral
Organs Removed
Plate 7

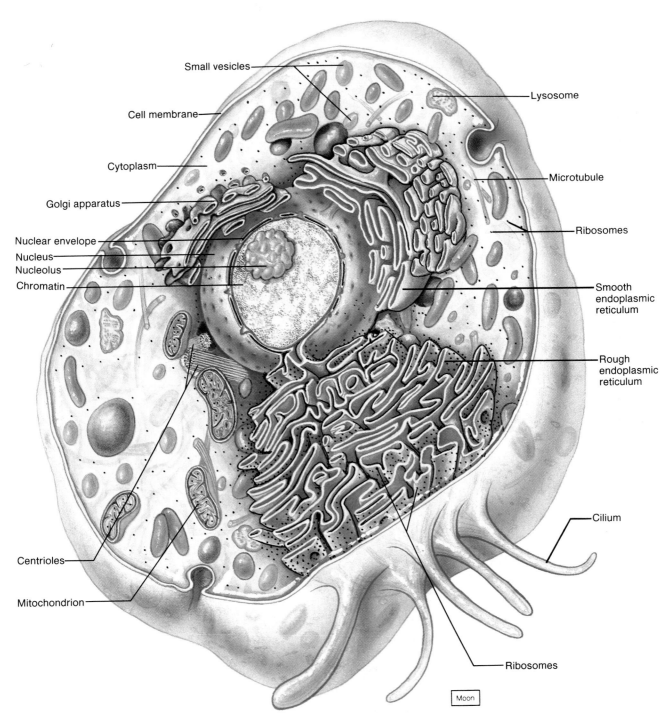

Small vesicles

Cell membrane

Cytoplasm

Golgi apparatus

Nuclear envelope

Nucleus

Nucleolus

Chromatin

Centrioles

Mitochondrion

Lysosome

Microtubule

Ribosomes

Smooth
endoplasmic
reticulum

Rough
endoplasmic
reticulum

Cilium

Ribosomes

Moon

Composite Cell
Figure 3.3

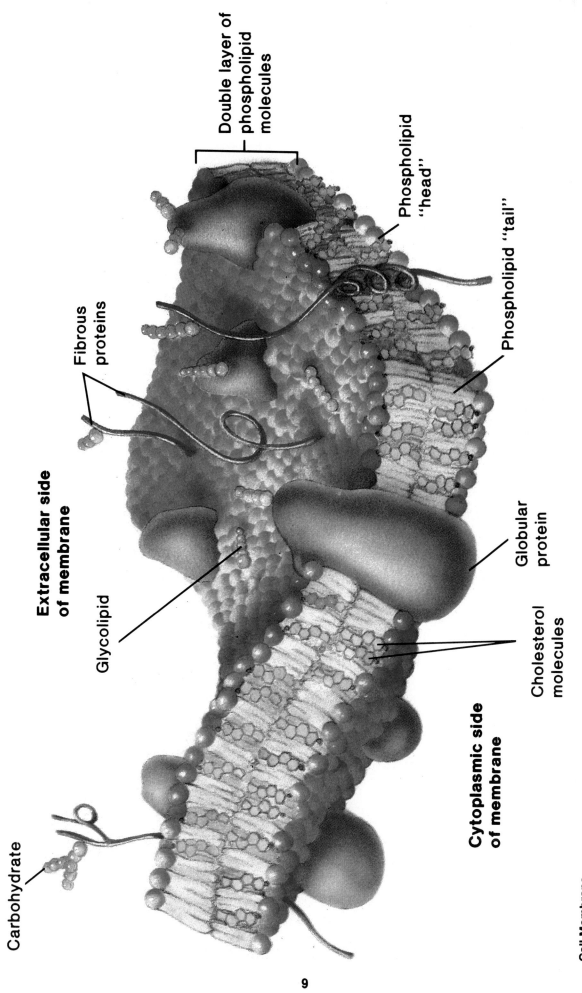

Double layer of phospholipid molecules

Phospholipid "head"

Phospholipid "tail"

Fibrous proteins

Extracellular side of membrane

Glycolipid

Globular protein

Cytoplasmic side of membrane

Cholesterol molecules

Carbohydrate

Cell Membrane
Figure 3.5

9

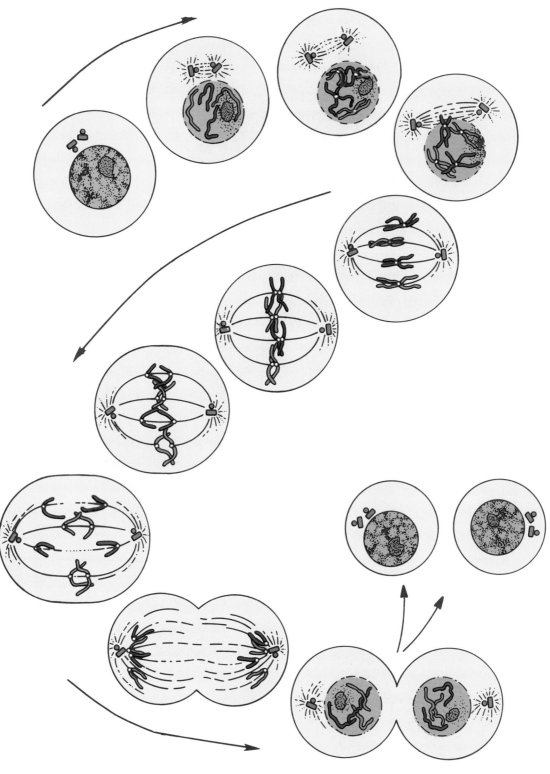

Mitosis
Figure 3.33

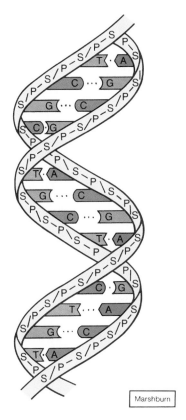

DNA Double Helix
Figure 4.19a

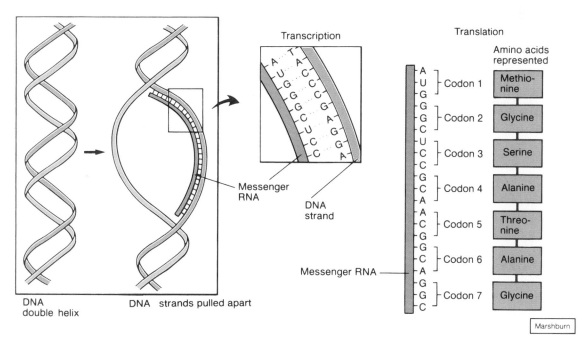

DNA–RNA Transcription and Translation
Figure 4.26

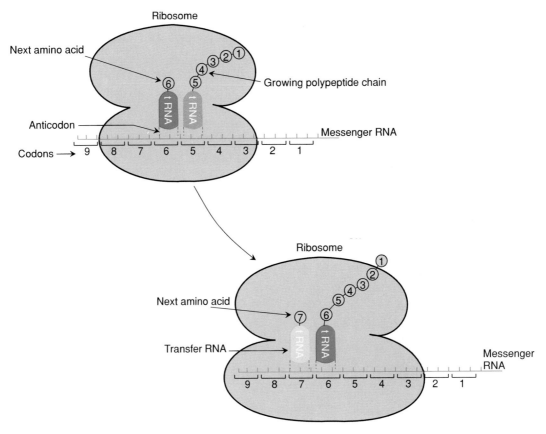

Protein Synthesis
Figure 4.27

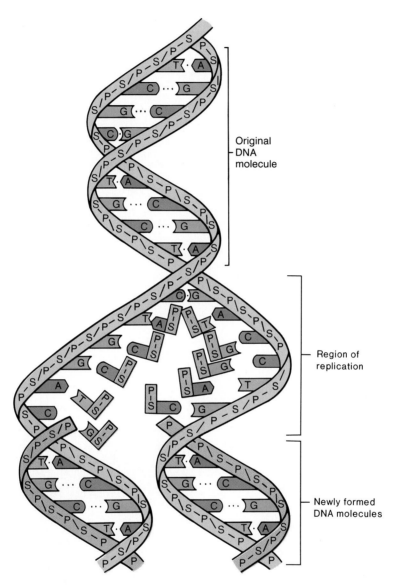

DNA Replication
Figure 4.28

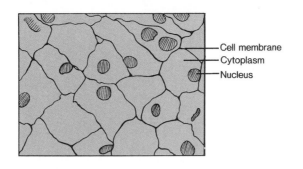

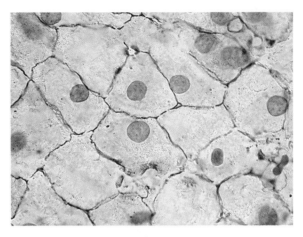

Simple Squamous Epithelium
Figure 5.1a–b

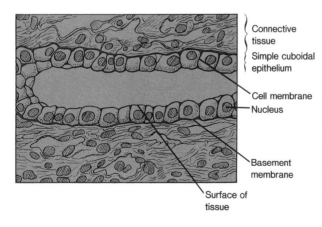

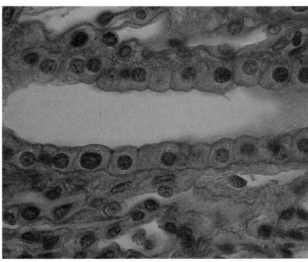

Simple Cuboidal Epithelium
Figure 5.2a–b

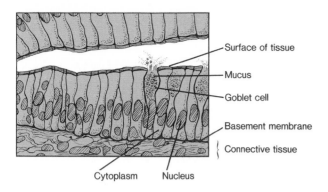

Surface of tissue

Mucus

Goblet cell

Basement membrane

{ Connective tissue

Cytoplasm   Nucleus

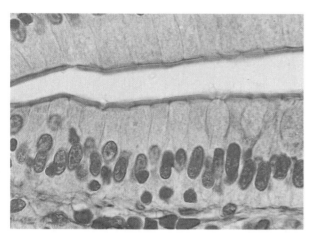

Simple Columnar Epithelium
Figure 5.3a–b

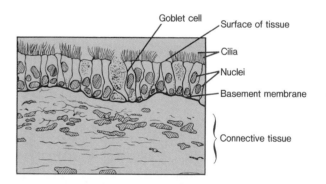

Goblet cell   Surface of tissue

Cilia

Nuclei

Basement membrane

} Connective tissue

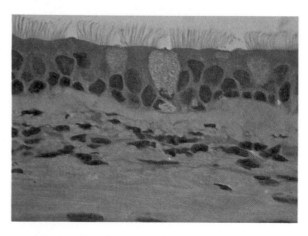

Pseudostratified Columnar Epithelium
Figure 5.5a–b

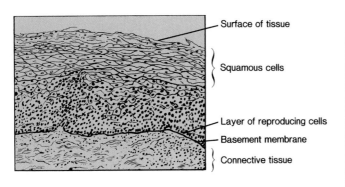

Surface of tissue

} Squamous cells

Layer of reproducing cells
Basement membrane
} Connective tissue

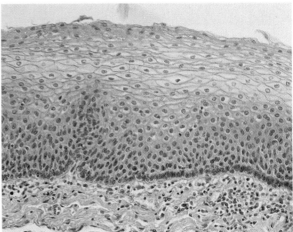

Stratified Squamous Epithelium
Figure 5.6a–b

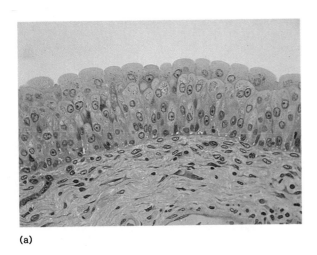

(a)

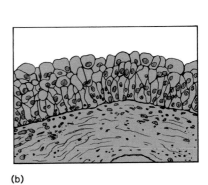

(b)

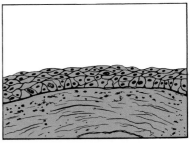

(c)

Transitional Epithelium, Contracted and Stretched
Figure 5.7a–c

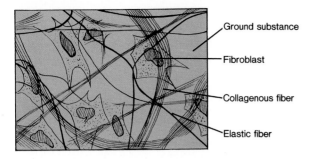

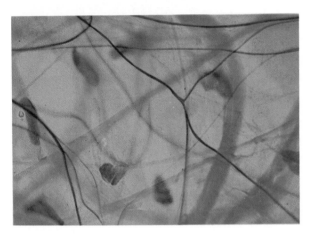

Loose Connective Tissue
Figure 5.15a–b

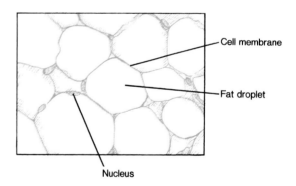

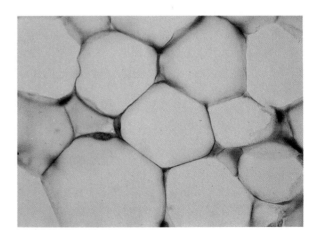

Adipose Tissue
Figure 5.16a–b

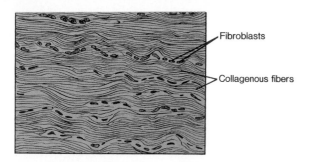

Fibroblasts

Collagenous fibers

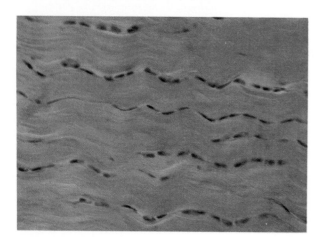

Fibrous Connective Tissue
Figure 5.17a–b

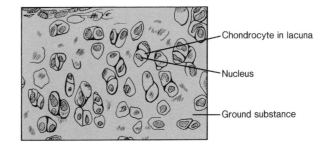

Chondrocyte in lacuna

Nucleus

Ground substance

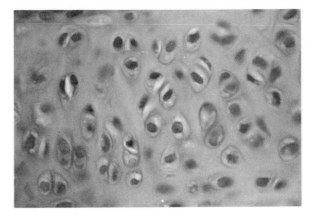

Hyaline Cartilage
Figure 5.20a–b

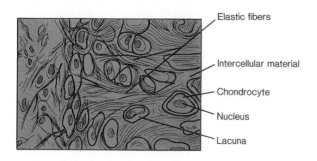

Elastic fibers

Intercellular material

Chondrocyte

Nucleus

Lacuna

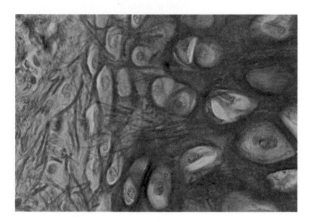

Elastic Cartilage
Figure 5.21a–b

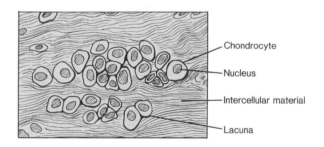

Chondrocyte

Nucleus

Intercellular material

Lacuna

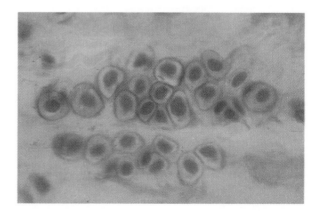

Fibrocartilage
Figure 5.22a–b

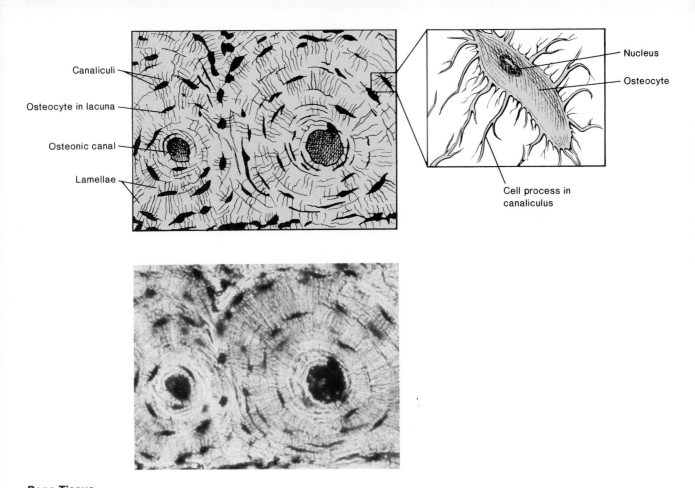

Canaliculi

Osteocyte in lacuna

Osteonic canal

Lamellae

Nucleus

Osteocyte

Cell process in canaliculus

Bone Tissue
Figure 5.23a–b

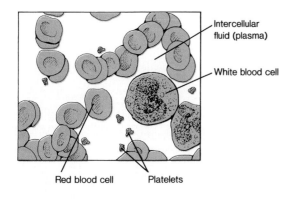

Intercellular fluid (plasma)

White blood cell

Red blood cell   Platelets

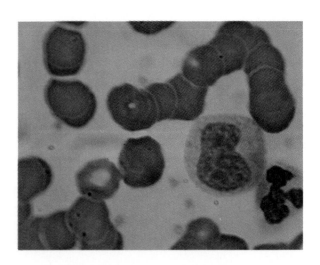

Blood Tissue
Figure 5.24a–b

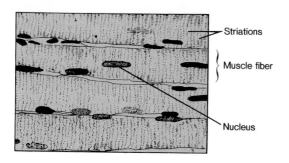

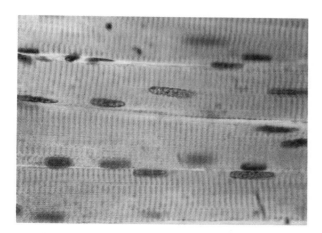

Skeletal Muscle Tissue
Figure 5.25a–b

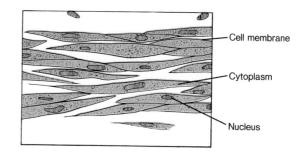

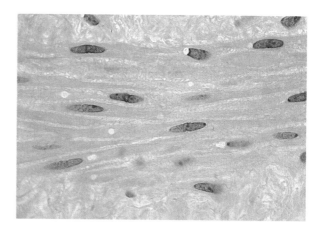

Smooth Muscle Tissue
Figure 5.26a–b

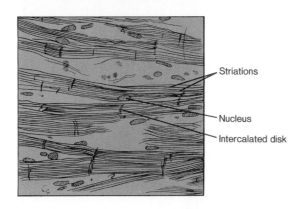

Striations

Nucleus

Intercalated disk

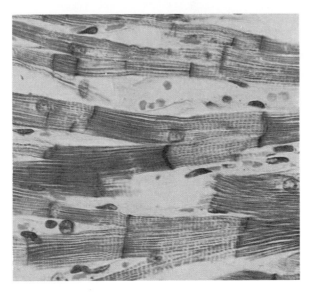

Cardiac Muscle Tissue
Figure 5.27a–b

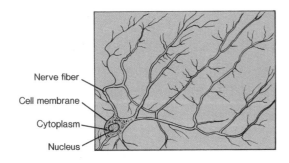

Nerve fiber

Cell membrane

Cytoplasm

Nucleus

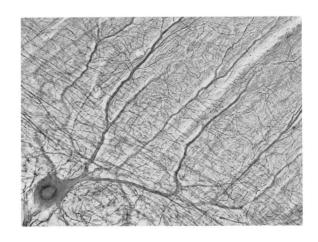

Nervous Tissue
Figure 5.28a–b

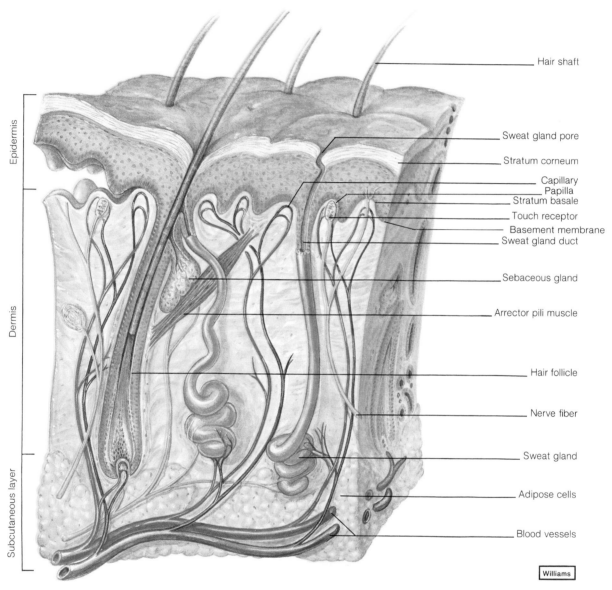

Epidermis

Dermis

Subcutaneous layer

Hair shaft

Sweat gland pore

Stratum corneum

Capillary
Papilla
Stratum basale
Touch receptor
Basement membrane
Sweat gland duct

Sebaceous gland

Arrector pili muscle

Hair follicle

Nerve fiber

Sweat gland

Adipose cells

Blood vessels

Williams

Skin Section
Figure 6.2

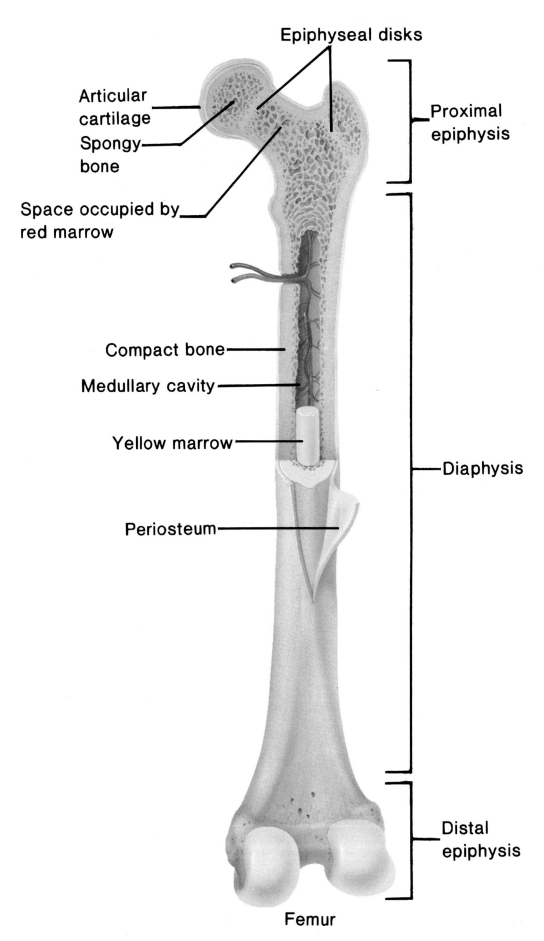

Epiphyseal disks

Articular
cartilage

Spongy
bone

Space occupied by
red marrow

Proximal
epiphysis

Compact bone

Medullary cavity

Yellow marrow

Periosteum

Diaphysis

Distal
epiphysis

Femur

Structure of a Long Bone
Figure 7.2

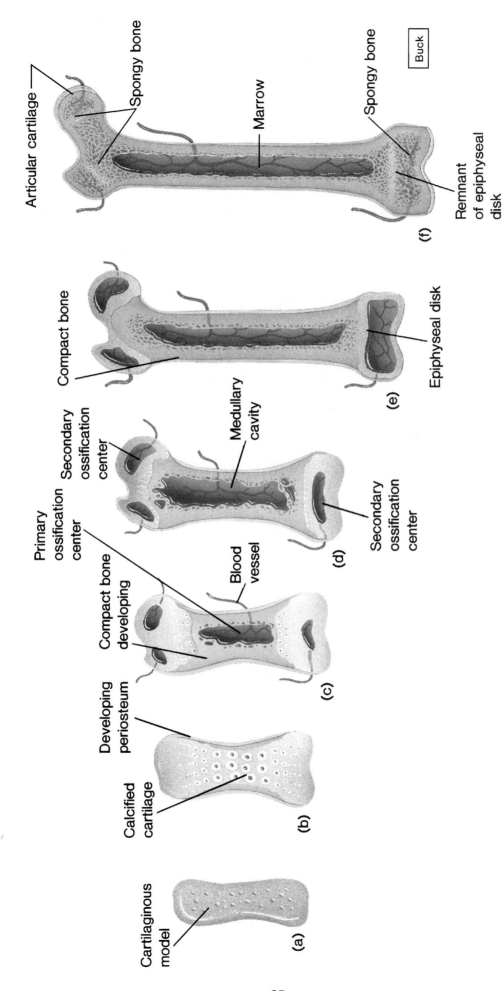

Cartilaginous model

(a)

Calcified cartilage

Developing periosteum

(b)

Blood vessel

Compact bone developing

Primary ossification center

(c)

Medullary cavity

Secondary ossification center

Secondary ossification center

(d)

Compact bone

Epiphyseal disk

(e)

Articular cartilage

Spongy bone

Marrow

Spongy bone

Remnant of epiphyseal disk

(f)

Buck

Development of a Long Bone
Figure 7.10

25

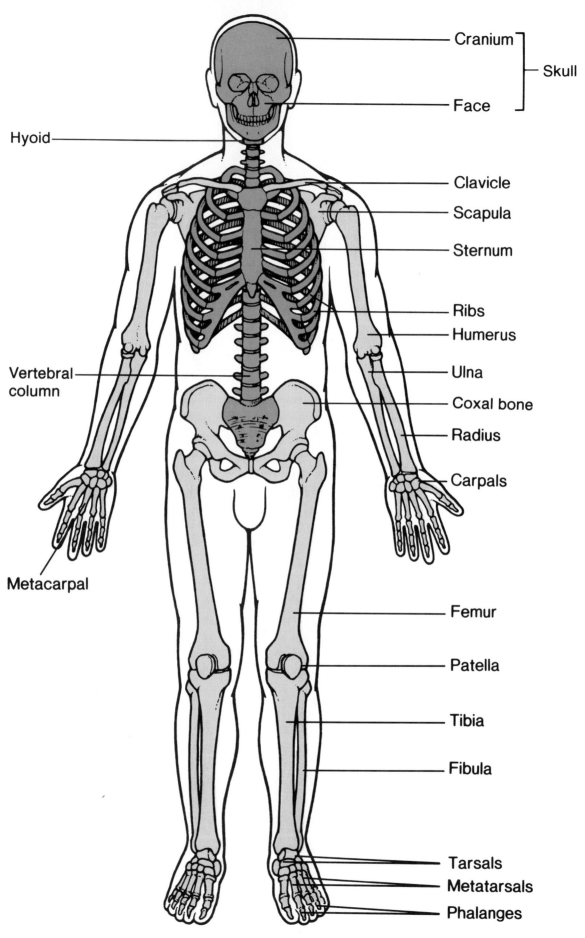

Cranium
Skull
Face

Hyoid

Clavicle

Scapula

Sternum

Ribs

Humerus

Ulna

Coxal bone

Radius

Carpals

Vertebral column

Metacarpal

Femur

Patella

Tibia

Fibula

Tarsals

Metatarsals

Phalanges

Human Skeleton, Anterior View
Figure 7.18a

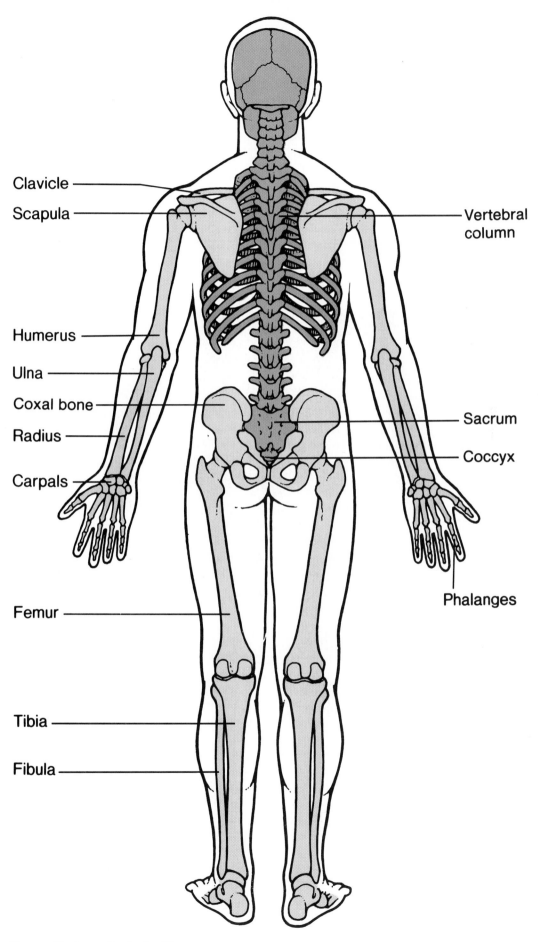

Clavicle

Scapula

Humerus

Ulna

Coxal bone

Radius

Carpals

Vertebral
column

Sacrum

Coccyx

Phalanges

Femur

Tibia

Fibula

Human Skeleton, Posterior View
Figure 7.18b

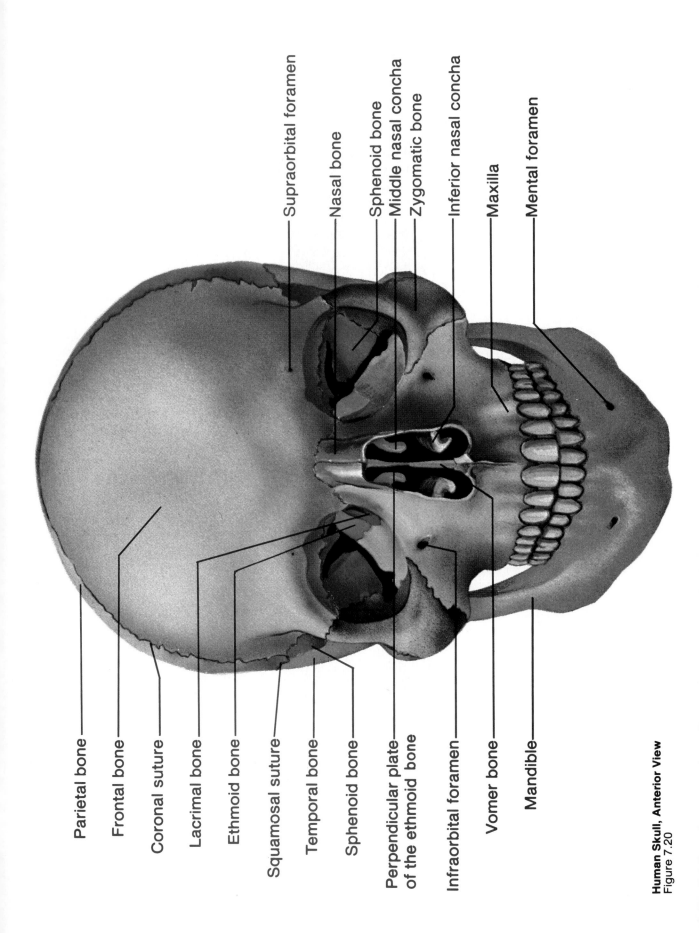

Parietal bone

Frontal bone

Coronal suture

Lacrimal bone

Ethmoid bone

Squamosal suture

Temporal bone

Sphenoid bone

Perpendicular plate
of the ethmoid bone

Infraorbital foramen

Vomer bone

Mandible

Supraorbital foramen

Nasal bone

Sphenoid bone

Middle nasal concha

Zygomatic bone

Inferior nasal concha

Maxilla

Mental foramen

Human Skull, Anterior View
Figure 7.20

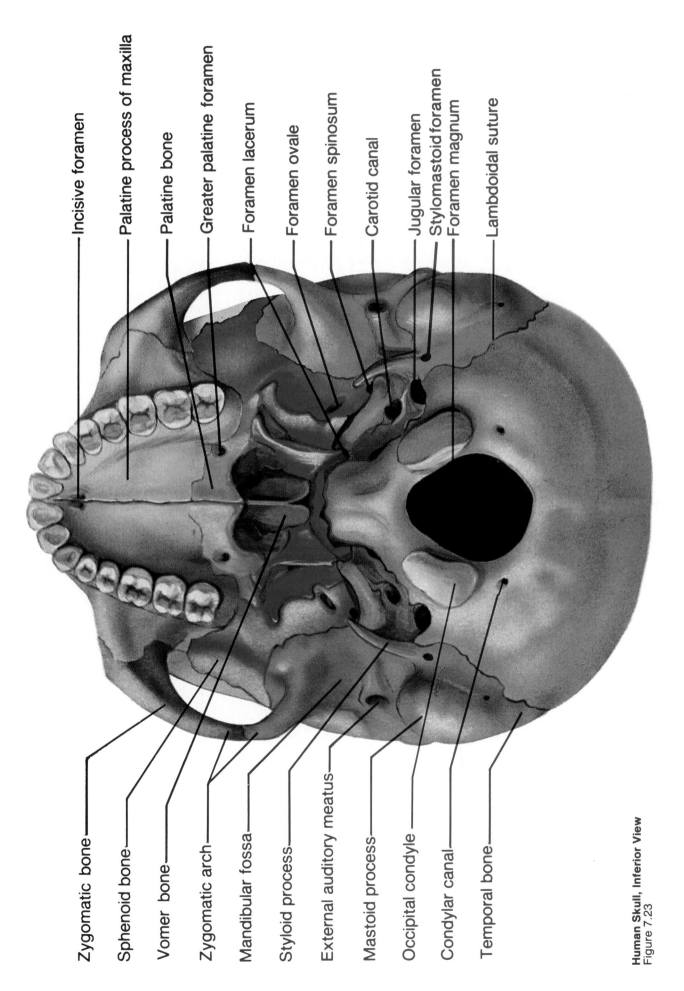

Incisive foramen

Palatine process of maxilla

Palatine bone

Greater palatine foramen

Foramen lacerum

Foramen ovale

Foramen spinosum

Carotid canal

Jugular foramen

Stylomastoid foramen

Foramen magnum

Lambdoidal suture

Zygomatic bone

Sphenoid bone

Vomer bone

Zygomatic arch

Mandibular fossa

Styloid process

External auditory meatus

Mastoid process

Occipital condyle

Condylar canal

Temporal bone

Human Skull, Inferior View
Figure 7.23

29

Crista galli

Cribriform plate (ethmoid bone)

Frontal bone

Sphenoid bone

Temporal bone

Foramen spinosum

Sella turcica

Parietal bone

Occipital bone

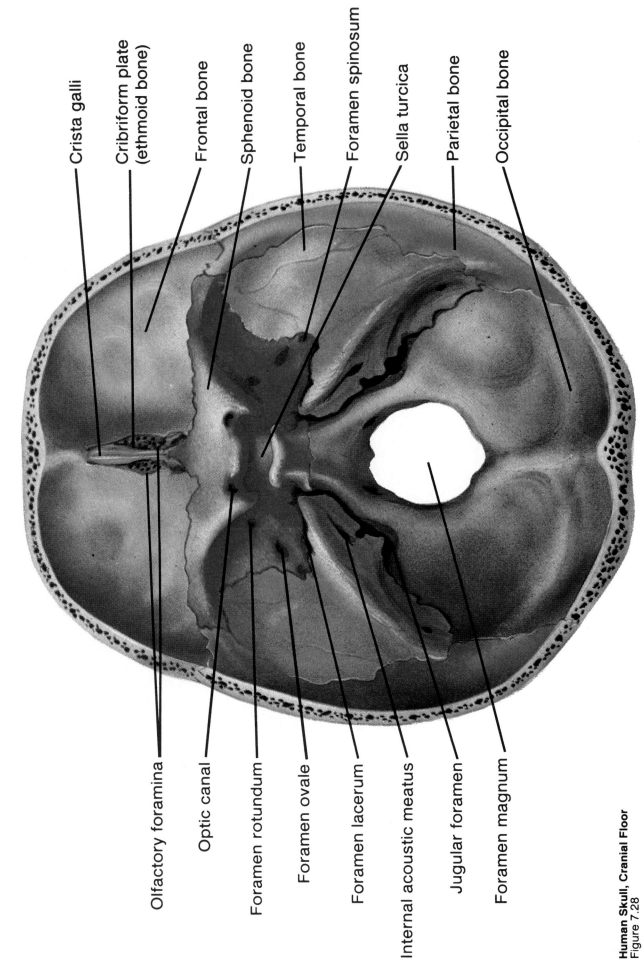

Olfactory foramina

Optic canal

Foramen rotundum

Foramen ovale

Foramen lacerum

Internal acoustic meatus

Jugular foramen

Foramen magnum

Human Skull, Cranial Floor
Figure 7.28

30

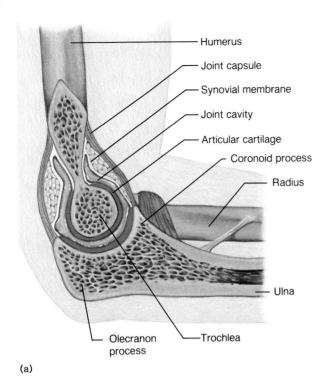

Humerus
Joint capsule
Synovial membrane
Joint cavity
Articular cartilage
Coronoid process
Radius
Ulna
Olecranon process
Trochlea

(a)

Elbow Joint
Figure 8.15a

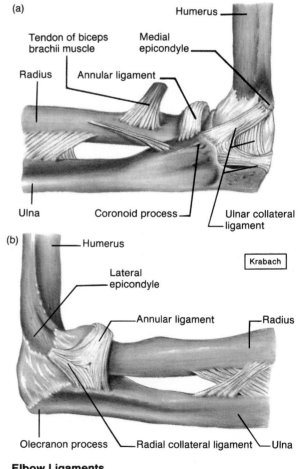

(a)

Humerus
Tendon of biceps brachii muscle
Medial epicondyle
Radius
Annular ligament
Ulna
Coronoid process
Ulnar collateral ligament

(b)

Humerus
Krabach
Lateral epicondyle
Annular ligament
Radius
Olecranon process
Radial collateral ligament
Ulna

Elbow Ligaments
Figure 8.16a–b

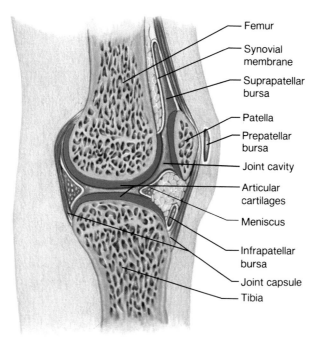

Femur

Synovial membrane

Suprapatellar bursa

Patella

Prepatellar bursa

Joint cavity

Articular cartilages

Meniscus

Infrapatellar bursa

Joint capsule

Tibia

(a)

Knee Joint
Figure 8.20a

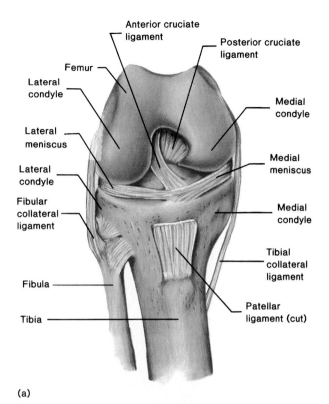

(a)

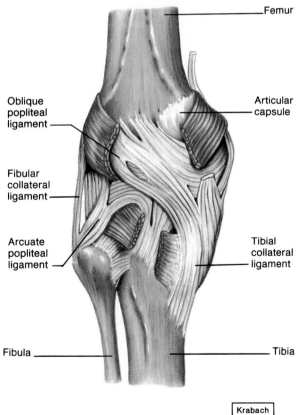

Krabach

(b)

Knee Ligaments
Figure 8.21a–b

33

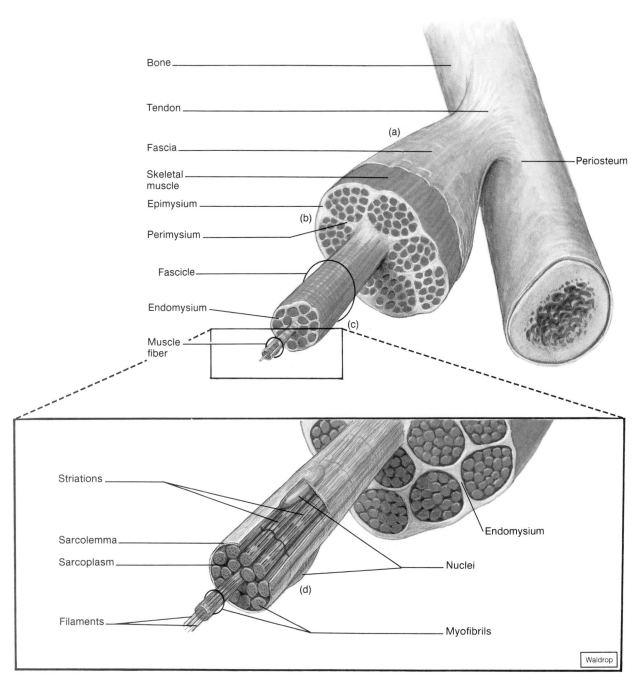

Skeletal Muscle Structure
Figure 9.2

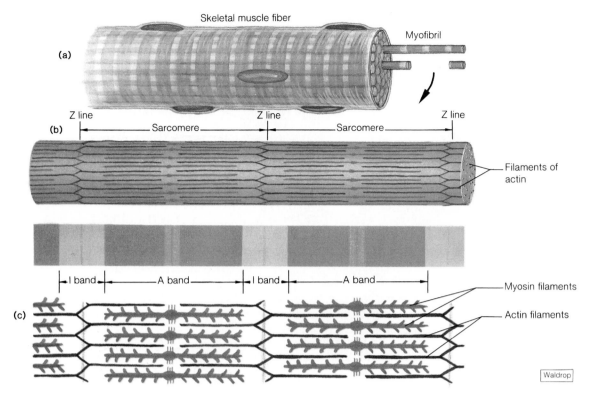

Skeletal Muscle, Fiber I
Figure 9.4

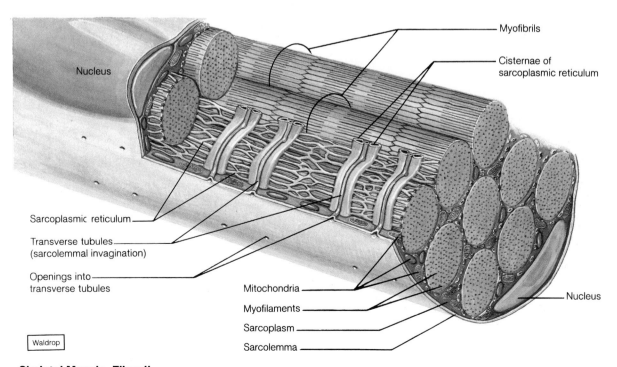

Skeletal Muscle, Fiber II
Figure 9.6

35

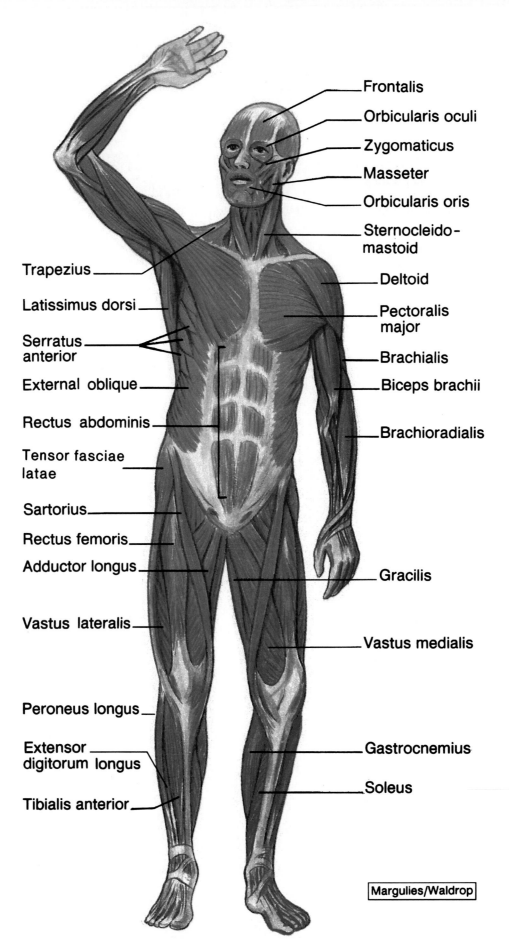

Frontalis

Orbicularis oculi

Zygomaticus

Masseter

Orbicularis oris

Sternocleido-
mastoid

Deltoid

Pectoralis
major

Brachialis

Biceps brachii

Brachioradialis

Trapezius

Latissimus dorsi

Serratus
anterior

External oblique

Rectus abdominis

Tensor fasciae
latae

Sartorius

Rectus femoris

Adductor longus

Gracilis

Vastus lateralis

Vastus medialis

Peroneus longus

Extensor
digitorum longus

Gastrocnemius

Soleus

Tibialis anterior

Margulies/Waldrop

Skeletal Muscles, Anterior View
Figure 9.20

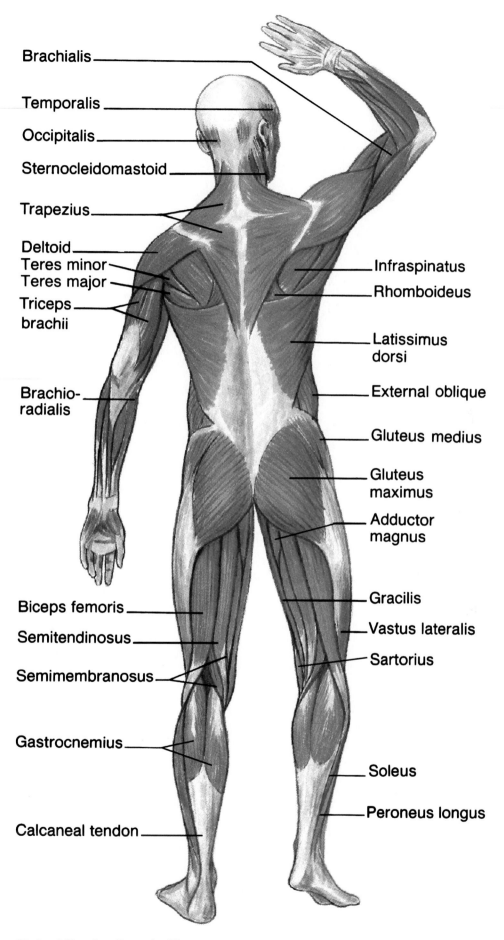

Brachialis

Temporalis

Occipitalis

Sternocleidomastoid

Trapezius

Deltoid
Teres minor
Teres major
Triceps
brachii

Brachio-
radialis

Infraspinatus

Rhomboideus

Latissimus
dorsi

External oblique

Gluteus medius

Gluteus
maximus

Adductor
magnus

Gracilis

Vastus lateralis

Sartorius

Biceps femoris

Semitendinosus

Semimembranosus

Gastrocnemius

Soleus

Peroneus longus

Calcaneal tendon

Skeletal Muscles, Posterior View
Figure 9.21

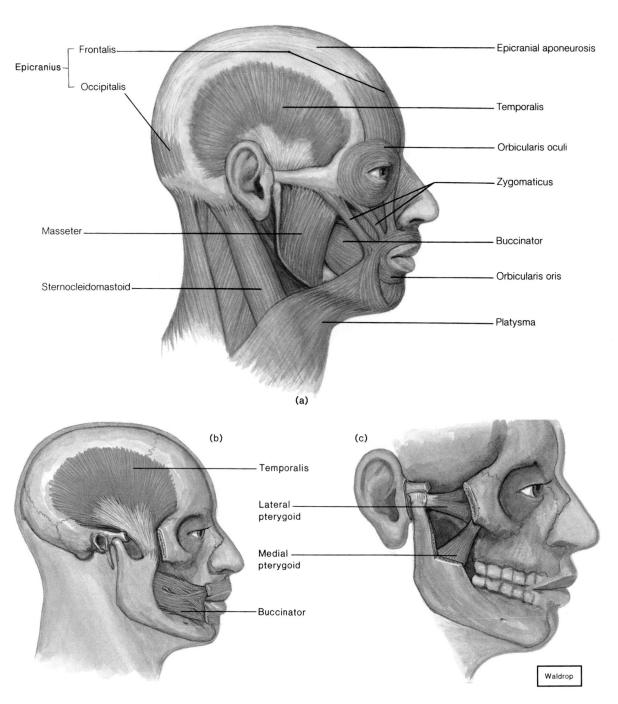

Epicranius
— Frontalis
— Occipitalis

Epicranial aponeurosis

Temporalis

Orbicularis oculi

Zygomaticus

Buccinator

Orbicularis oris

Platysma

Masseter

Sternocleidomastoid

(a)

(b)

(c)

Temporalis

Lateral pterygoid

Medial pterygoid

Buccinator

Waldrop

Muscles of Expression and Mastication
Figure 9.22a–c

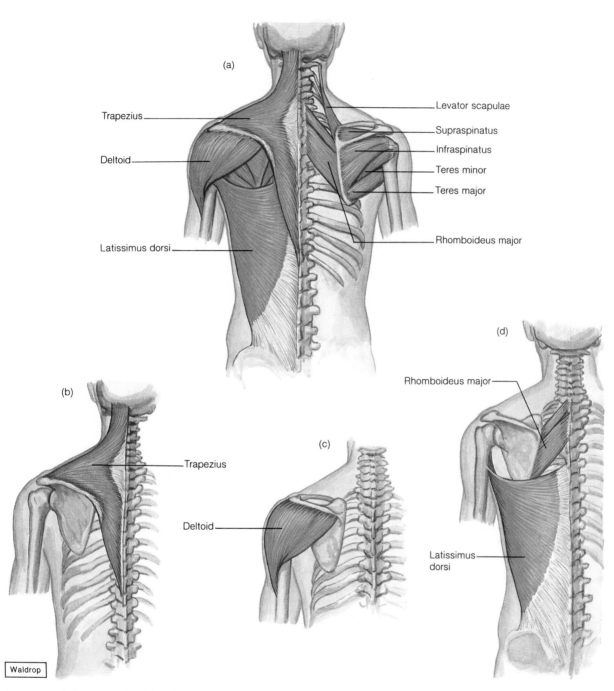

(a)

Trapezius

Deltoid

Latissimus dorsi

Levator scapulae

Supraspinatus

Infraspinatus

Teres minor

Teres major

Rhomboideus major

(b)

Trapezius

(c)

Deltoid

(d)

Rhomboideus major

Latissimus dorsi

Waldrop

Muscles of the Posterior Shoulder
Figure 9.24a–d

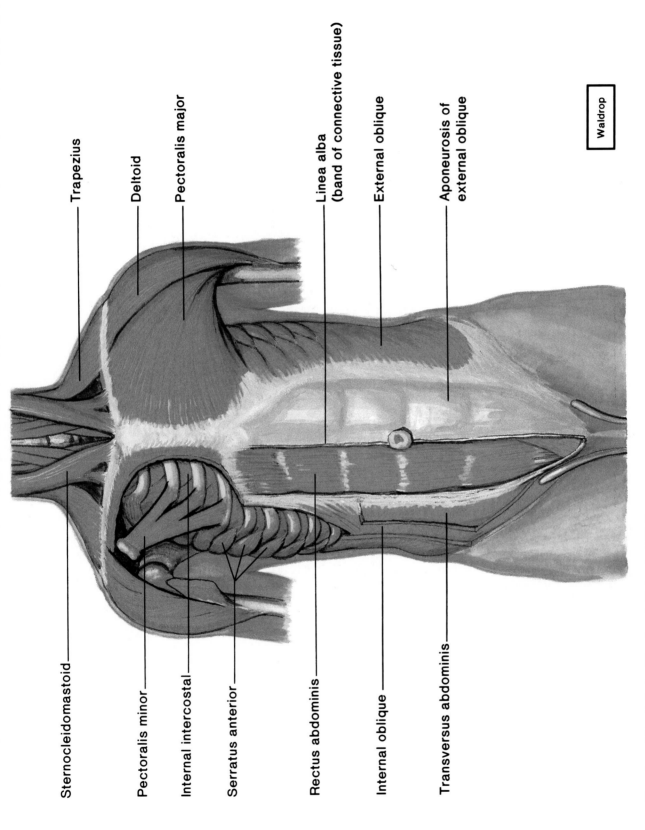

Sternocleidomastoid

Trapezius

Deltoid

Pectoralis major

Pectoralis minor

Internal intercostal

Serratus anterior

Linea alba
(band of connective tissue)

External oblique

Aponeurosis of
external oblique

Rectus abdominis

Internal oblique

Transversus abdominis

Waldrop

Muscles of the Anterior Chest and Abdominal Wall
Figure 9.25

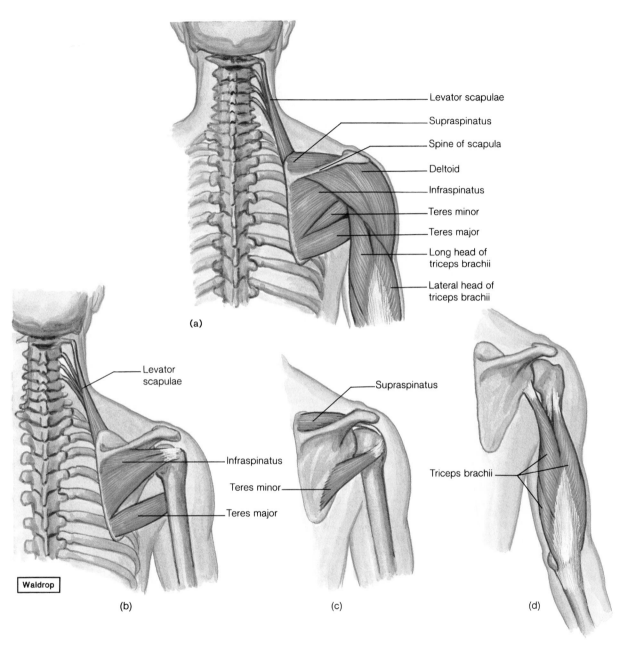

Levator scapulae
Supraspinatus
Spine of scapula
Deltoid
Infraspinatus
Teres minor
Teres major
Long head of triceps brachii
Lateral head of triceps brachii

(a)

Levator scapulae

Infraspinatus

Teres minor

Teres major

Waldrop

(b)

Supraspinatus

Teres minor

(c)

Triceps brachii

(d)

Muscles of the Scapula and Upper Arm
Figure 9.26a–d

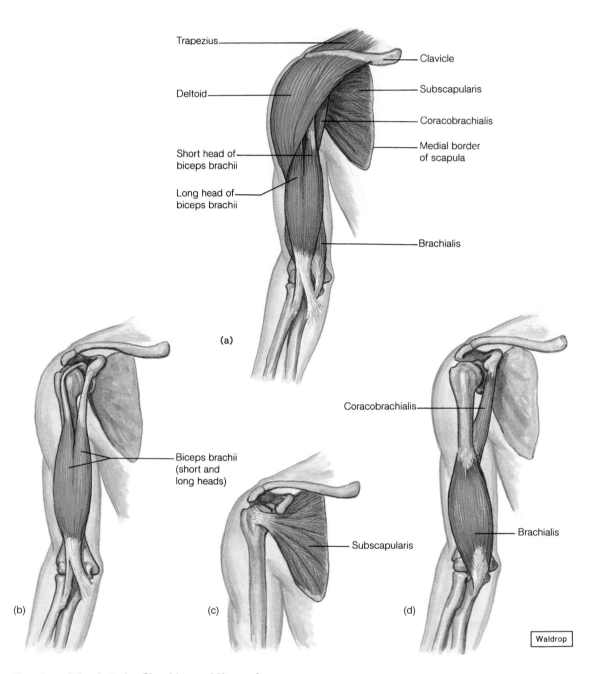

Muscles of the Anterior Shoulder and Upper Arm
Figure 9.28a–d

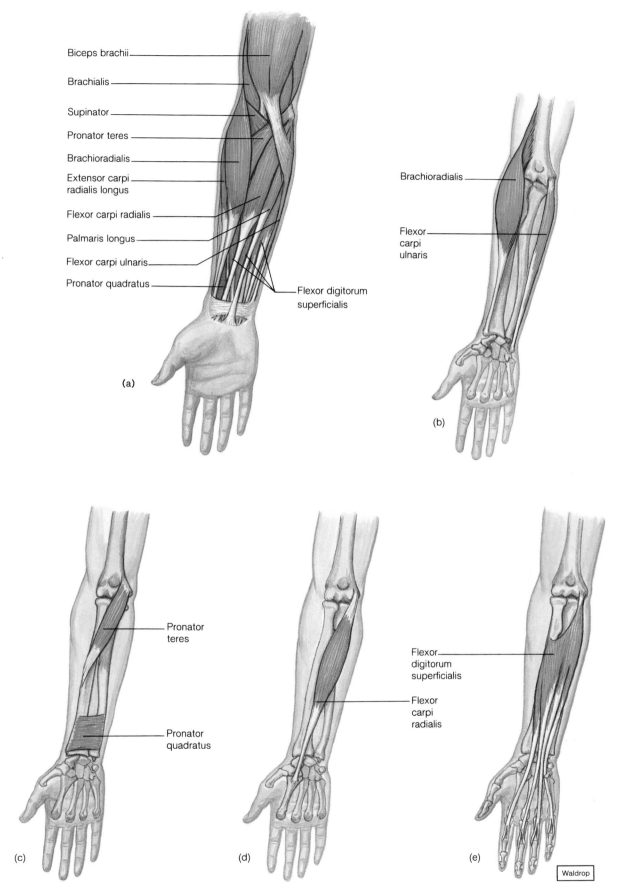

Biceps brachii

Brachialis

Supinator

Pronator teres

Brachioradialis

Extensor carpi radialis longus

Flexor carpi radialis

Palmaris longus

Flexor carpi ulnaris

Pronator quadratus

Flexor digitorum superficialis

(a)

Brachioradialis

Flexor carpi ulnaris

(b)

Pronator teres

Pronator quadratus

(c)

Flexor carpi radialis

(d)

Flexor digitorum superficialis

Flexor carpi radialis

(e)

Waldrop

Muscles of the Anterior Forearm
Figure 9.29a–e

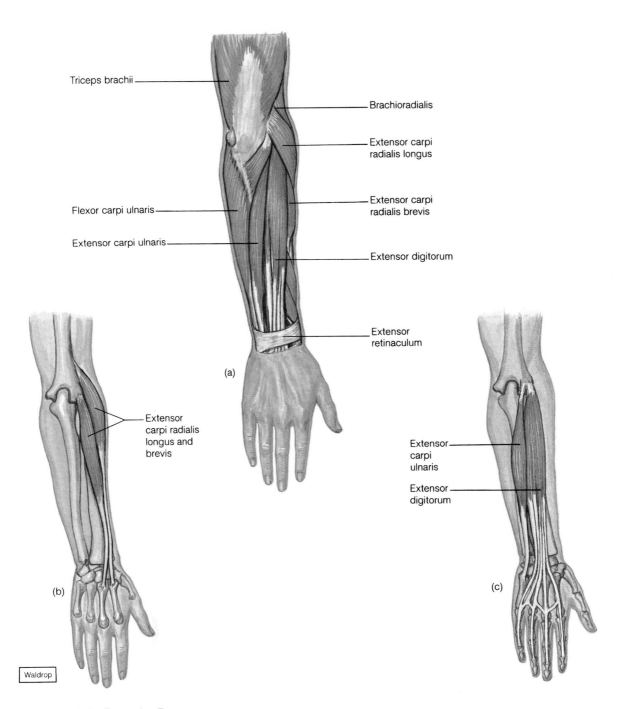

Triceps brachii

Brachioradialis

Extensor carpi radialis longus

Flexor carpi ulnaris

Extensor carpi ulnaris

Extensor carpi radialis brevis

Extensor digitorum

Extensor retinaculum

(a)

Extensor carpi radialis longus and brevis

(b)

Extensor carpi ulnaris

Extensor digitorum

(c)

Waldrop

Muscles of the Posterior Forearm
Figure 9.30a–c

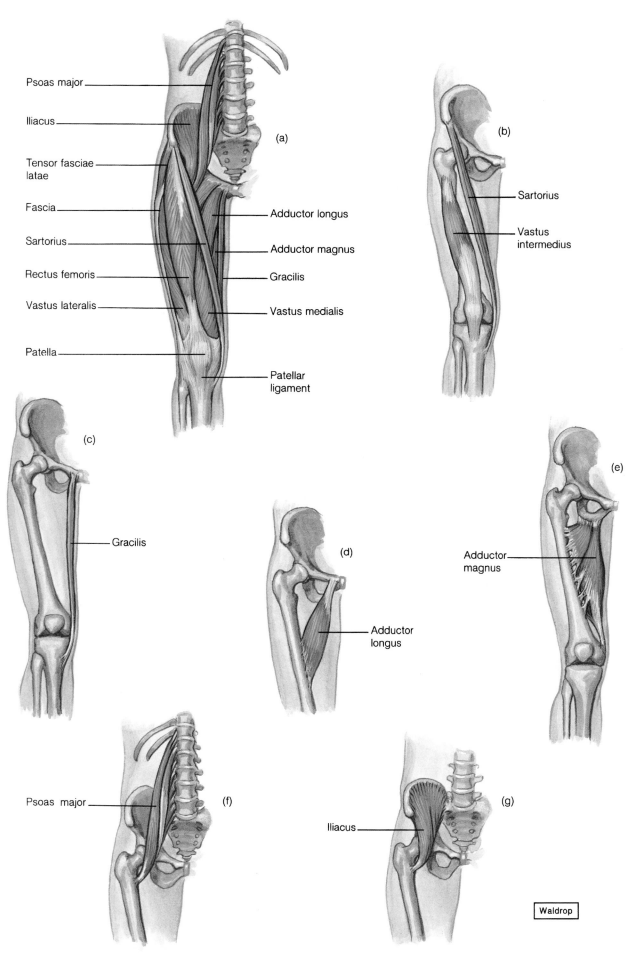

Psoas major

Iliacus

Tensor fasciae latae

Fascia

Sartorius

Rectus femoris

Vastus lateralis

Patella

(a)

Adductor longus

Adductor magnus

Gracilis

Vastus medialis

Patellar ligament

(b)

Sartorius

Vastus intermedius

(c)

Gracilis

(d)

Adductor longus

(e)

Adductor magnus

(f)

Psoas major

(g)

Iliacus

Waldrop

Muscles of the Anterior Thigh
Figure 9.34a–g

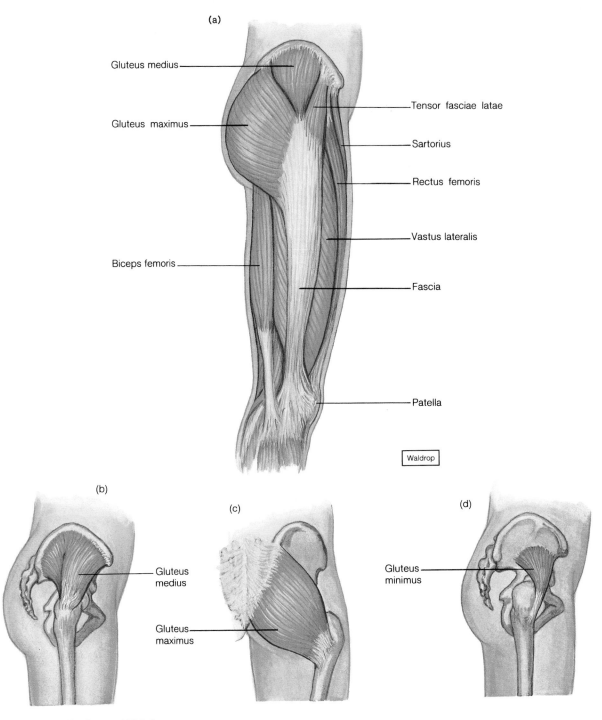

(a)

Gluteus medius

Gluteus maximus

Biceps femoris

Tensor fasciae latae

Sartorius

Rectus femoris

Vastus lateralis

Fascia

Patella

Waldrop

(b)

Gluteus medius

(c)

Gluteus maximus

(d)

Gluteus minimus

Muscles of the Lateral Thigh
Figure 9.35a–d

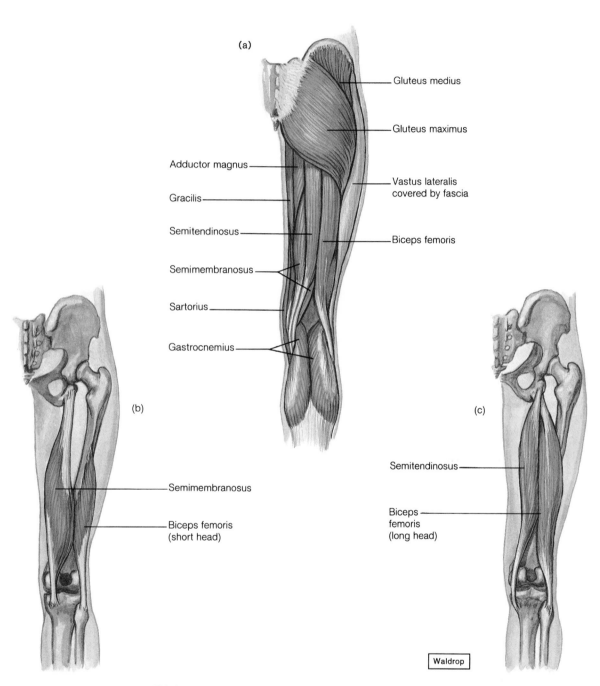

(a)

Gluteus medius

Gluteus maximus

Adductor magnus

Gracilis

Semitendinosus

Semimembranosus

Sartorius

Gastrocnemius

Vastus lateralis
covered by fascia

Biceps femoris

(b)

Semimembranosus

Biceps femoris
(short head)

(c)

Semitendinosus

Biceps
femoris
(long head)

Waldrop

Muscles of the Posterior Thigh
Figure 9.36a–c

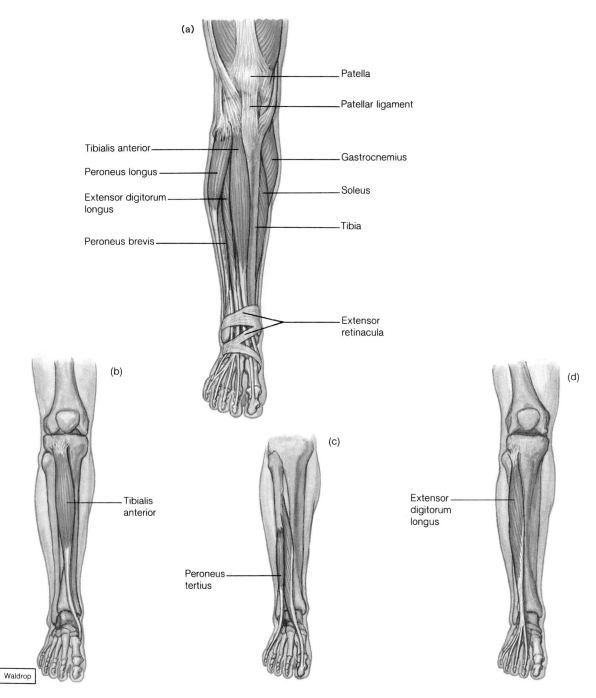

(a)

Patella

Patellar ligament

Tibialis anterior

Peroneus longus

Extensor digitorum longus

Peroneus brevis

Gastrocnemius

Soleus

Tibia

Extensor retinacula

(b)

Tibialis anterior

(c)

Peroneus tertius

(d)

Extensor digitorum longus

Waldrop

Muscles of the Anterior Lower Leg
Figure 9.38a–d

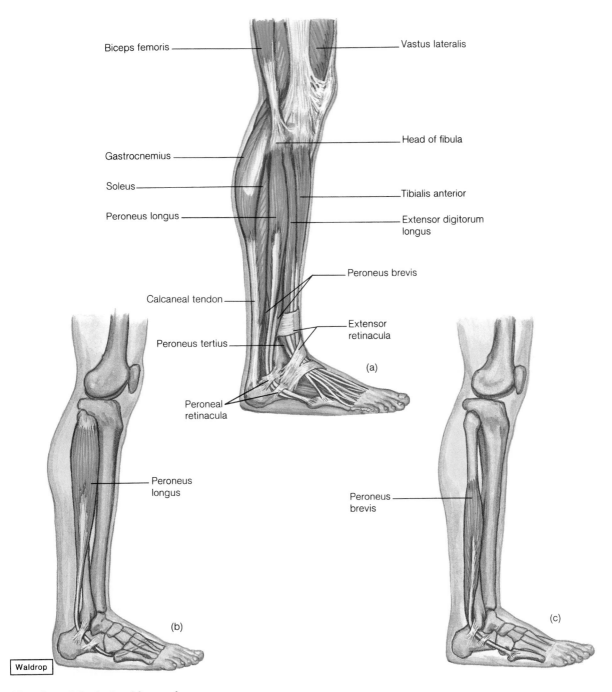

Biceps femoris

Vastus lateralis

Head of fibula

Gastrocnemius

Soleus

Tibialis anterior

Peroneus longus

Extensor digitorum longus

Peroneus brevis

Calcaneal tendon

Extensor retinacula

Peroneus tertius

(a)

Peroneal retinacula

Peroneus longus

Peroneus brevis

(b)

(c)

Waldrop

Muscles of the Lateral Lower Leg
Figure 9.39a–c

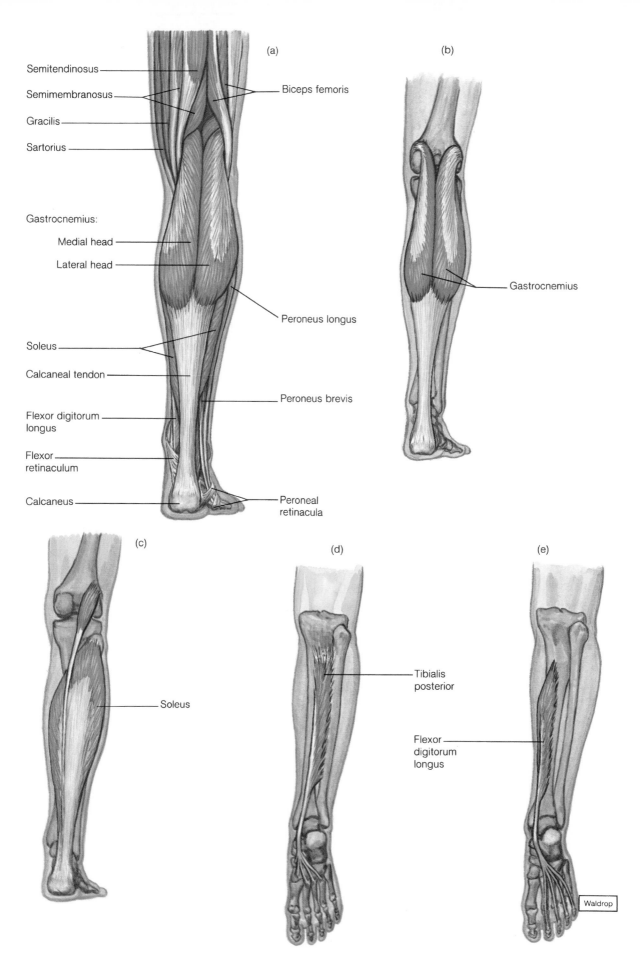

(a)

Semitendinosus

Semimembranosus

Gracilis

Sartorius

Biceps femoris

Gastrocnemius:

Medial head

Lateral head

Peroneus longus

Soleus

Calcaneal tendon

Peroneus brevis

Flexor digitorum longus

Flexor retinaculum

Calcaneus

Peroneal retinacula

(b)

Gastrocnemius

(c)

Soleus

(d)

Tibialis posterior

(e)

Flexor digitorum longus

Waldrop

Muscles of the Posterior Lower Leg
Figure 9.40a–e

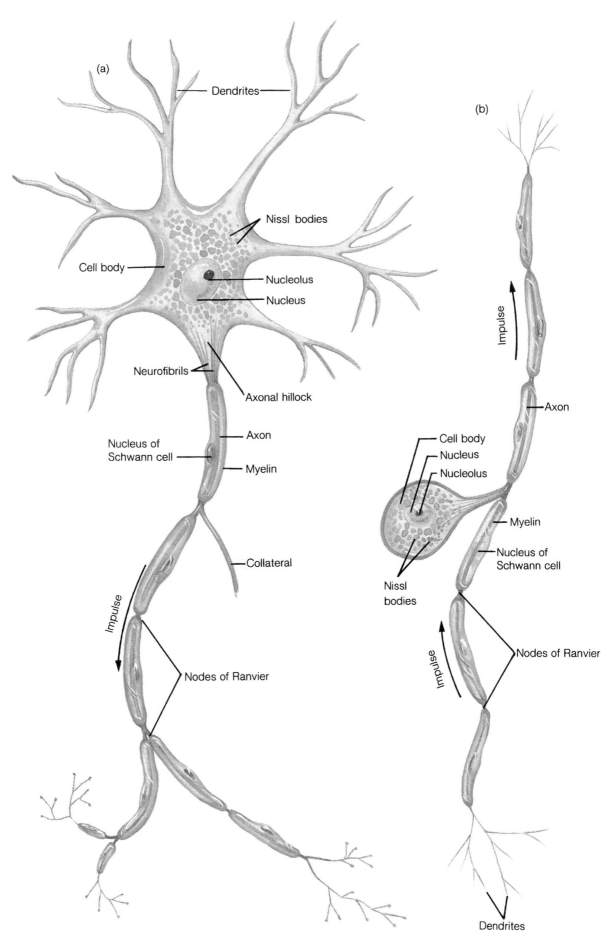

(a)

Dendrites

Nissl bodies

Cell body

Nucleolus

Nucleus

Neurofibrils

Axonal hillock

Axon

Nucleus of
Schwann cell

Myelin

Impulse

Collateral

Nodes of Ranvier

(b)

Impulse

Axon

Cell body

Nucleus

Nucleolus

Myelin

Nucleus of
Schwann cell

Nissl
bodies

Impulse

Nodes of Ranvier

Dendrites

Motor and Sensory Neurons
Figure 10.3a–b

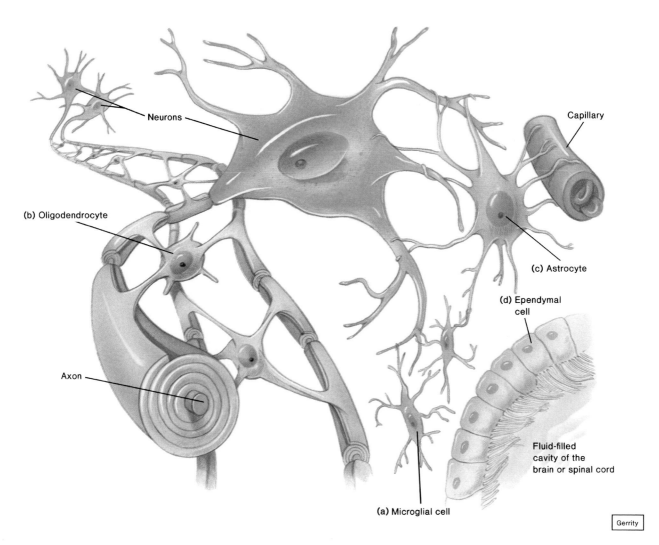

Neurons

Capillary

(b) Oligodendrocyte

(c) Astrocyte

(d) Ependymal cell

Axon

Fluid-filled cavity of the brain or spinal cord

(a) Microglial cell

Gerrity

Neuroglial Cells
Figure 10.6.

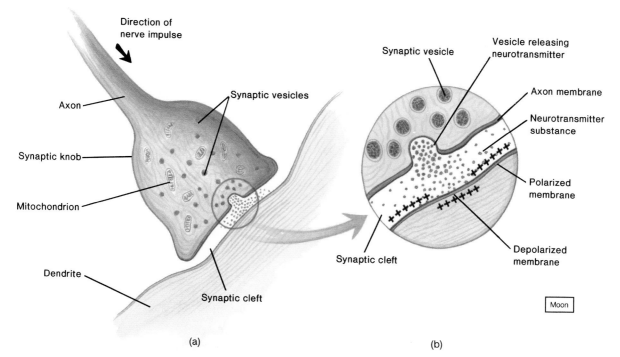

Synaptic Knob
Figure 10.19

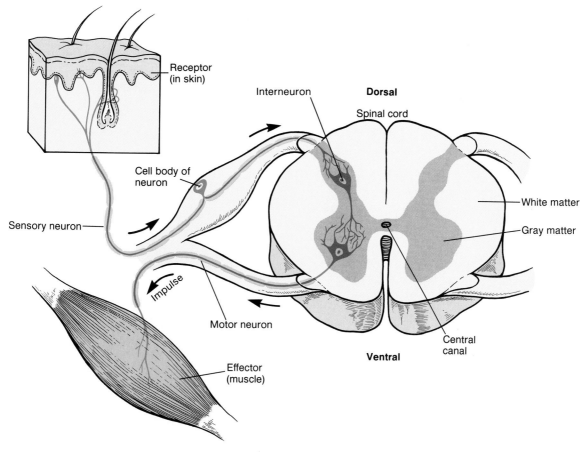

Reflex Arc
Figure 10.25

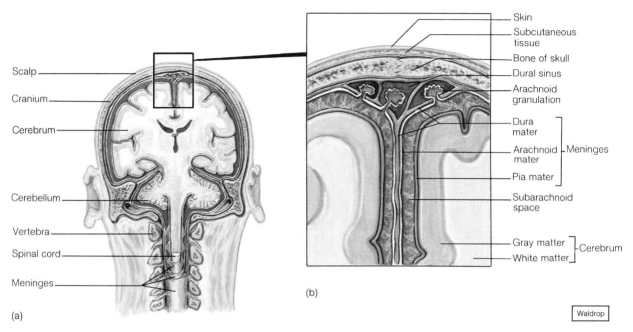

Scalp

Cranium

Cerebrum

Cerebellum

Vertebra

Spinal cord

Meninges

(a)

Skin

Subcutaneous tissue

Bone of skull

Dural sinus

Arachnoid granulation

Dura mater

Arachnoid mater ⎤ Meninges

Pia mater

Subarachnoid space

Gray matter ⎤ Cerebrum

White matter

(b)

Waldrop

Meninges
Figure 11.1

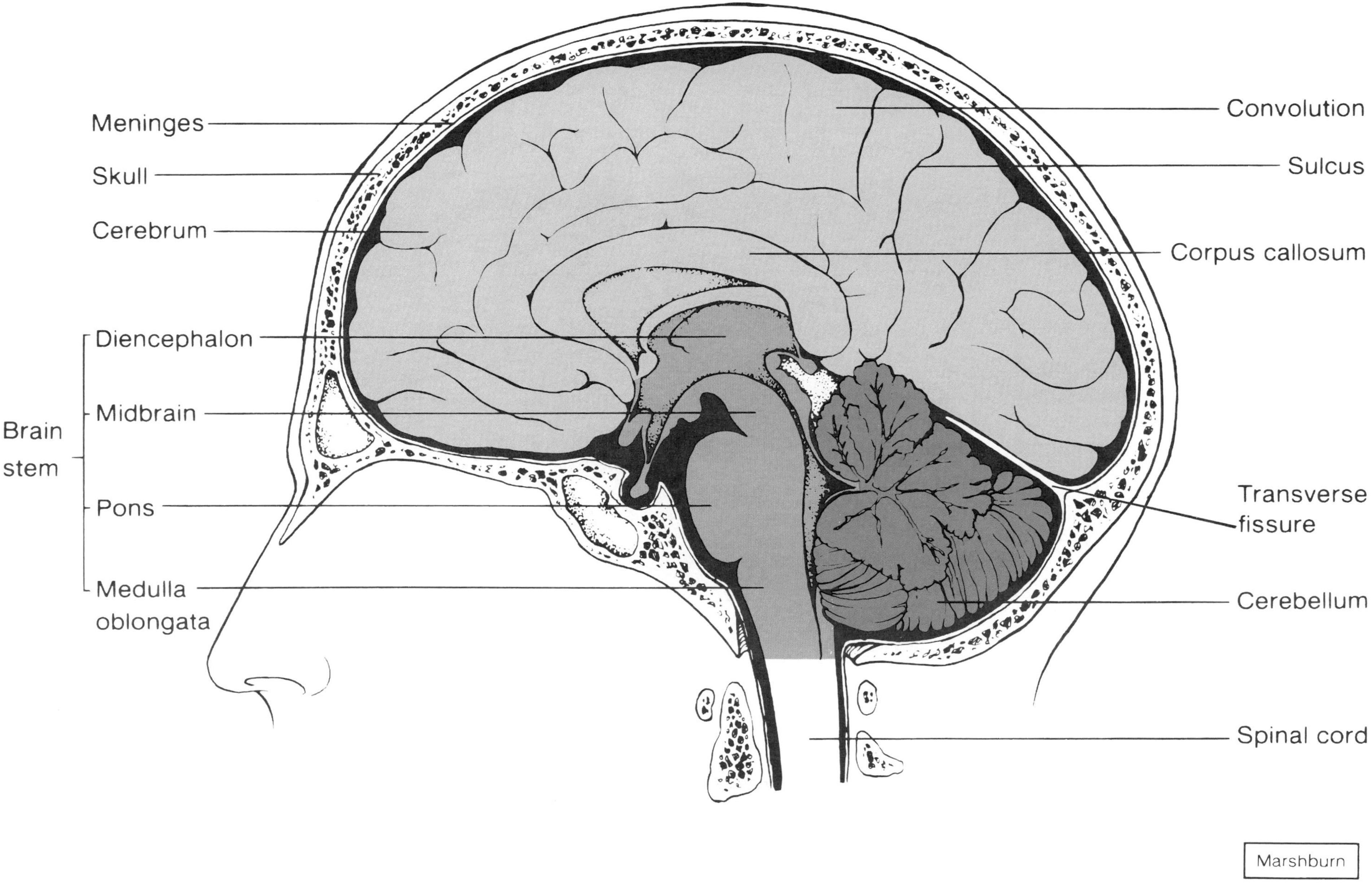

Convolution

Sulcus

Corpus callosum

Transverse
fissure

Cerebellum

Spinal cord

Marshburn

Meninges

Skull

Cerebrum

Diencephalon

Midbrain

Pons

Medulla
oblongata

Brain
stem

55

Brain, Sagittal Section
Figure 11.10

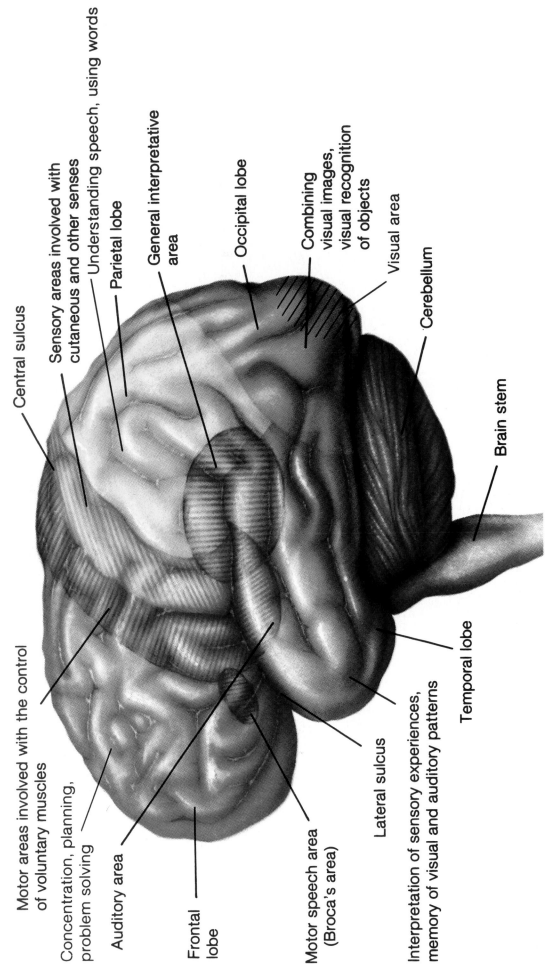

Central sulcus

Sensory areas involved with cutaneous and other senses

Understanding speech, using words

Parietal lobe

General interpretative area

Occipital lobe

Combining visual images, visual recognition of objects

Visual area

Cerebellum

Brain stem

Motor areas involved with the control of voluntary muscles

Concentration, planning, problem solving

Auditory area

Frontal lobe

Motor speech area (Broca's area)

Lateral sulcus

Interpretation of sensory experiences, memory of visual and auditory patterns

Temporal lobe

Sensory, Motor, and Association Areas
Figure 11.12

56

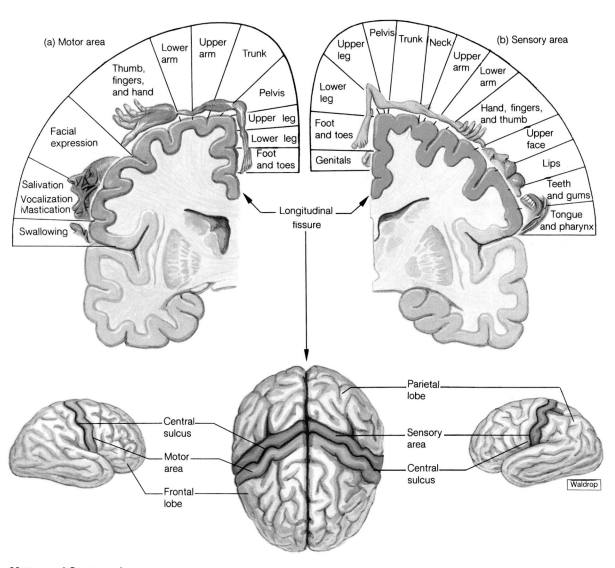

Motor and Sensory Areas
Figure 11.13

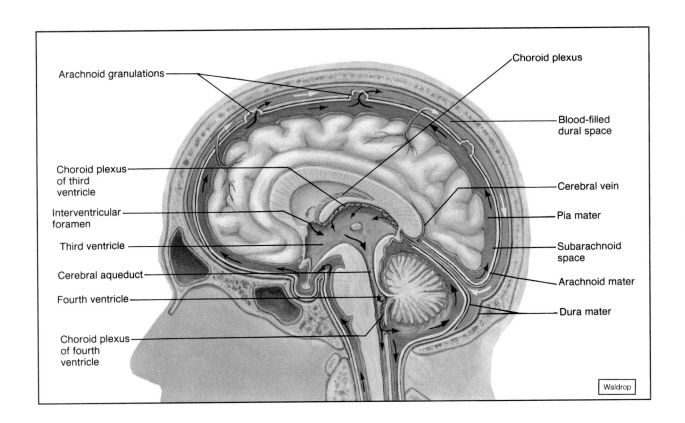

Arachnoid granulations

Choroid plexus

Blood-filled
dural space

Choroid plexus
of third
ventricle

Cerebral vein

Interventricular
foramen

Pia mater

Third ventricle

Subarachnoid
space

Cerebral aqueduct

Arachnoid mater

Fourth ventricle

Dura mater

Choroid plexus
of fourth
ventricle

Waldrop

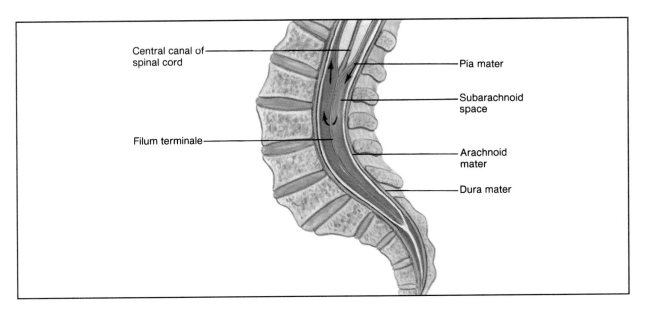

Central canal of
spinal cord

Pia mater

Subarachnoid
space

Filum terminale

Arachnoid
mater

Dura mater

Cerebrospinal Fluid Circulation
Figure 11.16

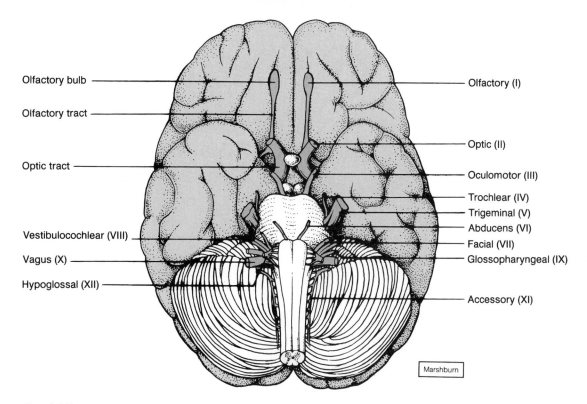

Cranial Nerves
Figure 11.26

Olfactory bulb
Olfactory tract
Optic tract
Vestibulocochlear (VIII)
Vagus (X)
Hypoglossal (XII)

Olfactory (I)
Optic (II)
Oculomotor (III)
Trochlear (IV)
Trigeminal (V)
Abducens (VI)
Facial (VII)
Glossopharyngeal (IX)
Accessory (XI)

Marshburn

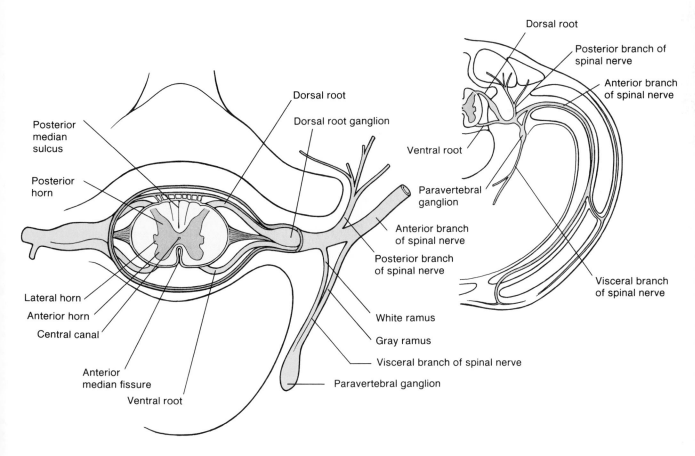

Posterior median sulcus
Posterior horn
Lateral horn
Anterior horn
Central canal
Anterior median fissure
Ventral root

Dorsal root
Dorsal root ganglion
Anterior branch of spinal nerve
Posterior branch of spinal nerve
White ramus
Gray ramus
Visceral branch of spinal nerve
Paravertebral ganglion

Dorsal root
Posterior branch of spinal nerve
Anterior branch of spinal nerve
Ventral root
Paravertebral ganglion
Visceral branch of spinal nerve

Spinal Nerve Branches
Figure 11.32

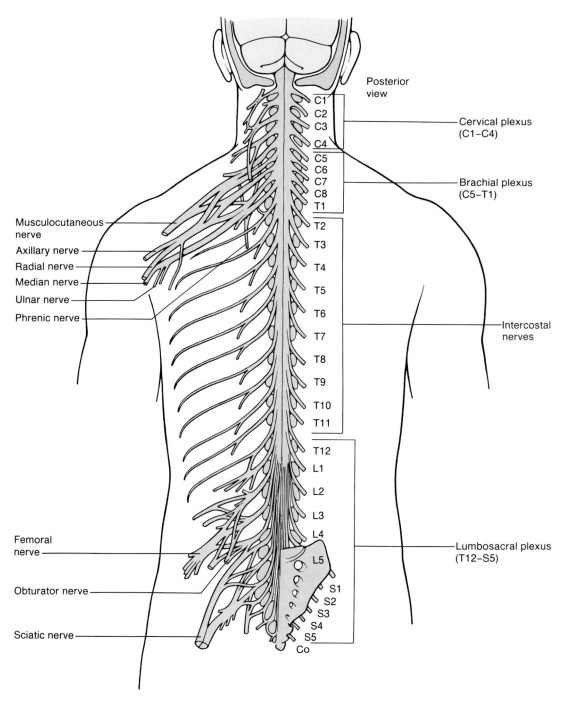

Posterior view

Cervical plexus (C1–C4)

Brachial plexus (C5–T1)

Intercostal nerves

Lumbosacral plexus (T12–S5)

Musculocutaneous nerve

Axillary nerve

Radial nerve

Median nerve

Ulnar nerve

Phrenic nerve

Femoral nerve

Obturator nerve

Sciatic nerve

C1
C2
C3
C4
C5
C6
C7
C8
T1
T2
T3
T4
T5
T6
T7
T8
T9
T10
T11
T12
L1
L2
L3
L4
L5
S1
S2
S3
S4
S5
Co

Spinal Nerves and Plexuses
Figure 11.33

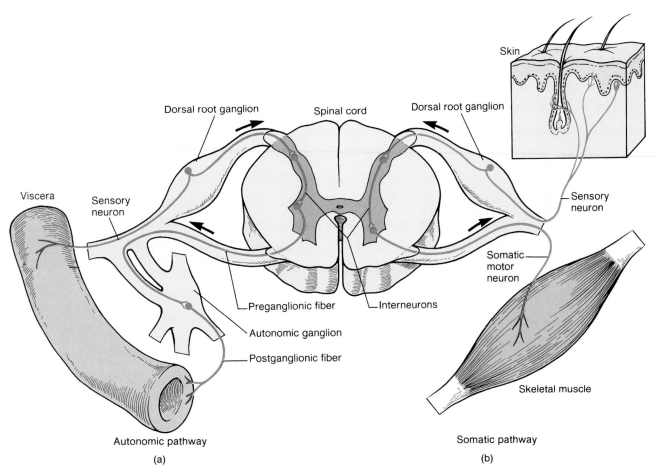

Dorsal root ganglion

Spinal cord

Dorsal root ganglion

Skin

Viscera

Sensory neuron

Sensory neuron

Somatic motor neuron

Preganglionic fiber

Interneurons

Autonomic ganglion

Postganglionic fiber

Skeletal muscle

Autonomic pathway

(a)

Somatic pathway

(b)

Autonomic and Somatic Nerve Pathways
Figure 11.36

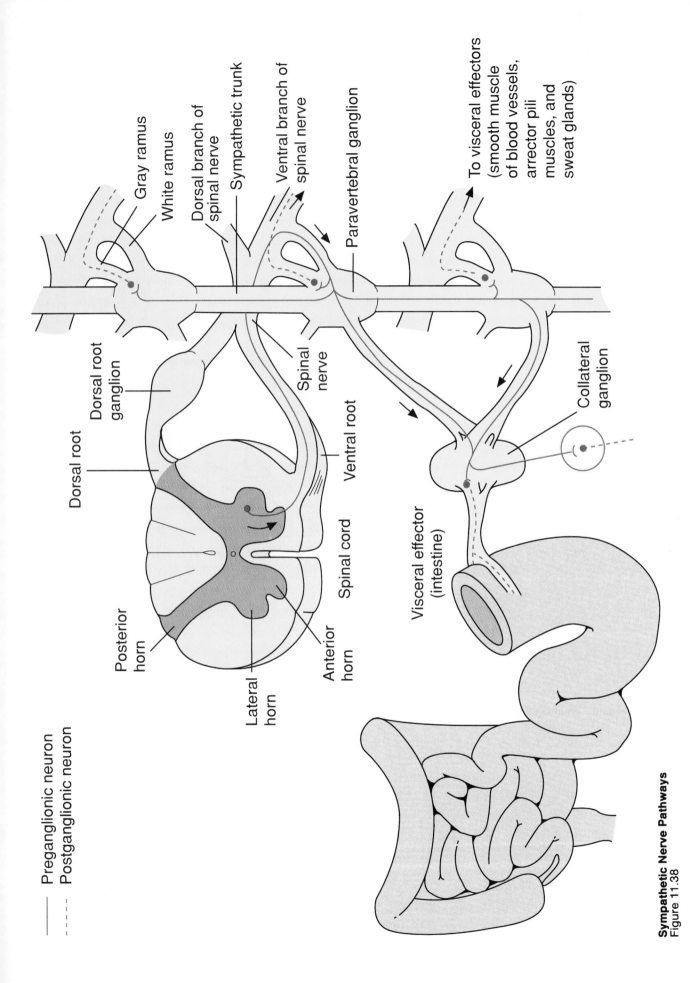

Sympathetic Nerve Pathways
Figure 11.38

Gray ramus
White ramus
Dorsal branch of spinal nerve
Sympathetic trunk
Ventral branch of spinal nerve
Paravertebral ganglion

To visceral effectors (smooth muscle of blood vessels, arrector pili muscles, and sweat glands)

Dorsal root ganglion
Dorsal root
Spinal nerve
Ventral root
Spinal cord
Anterior horn
Lateral horn
Posterior horn

Visceral effector (intestine)
Collateral ganglion

Preganglionic neuron
Postganglionic neuron

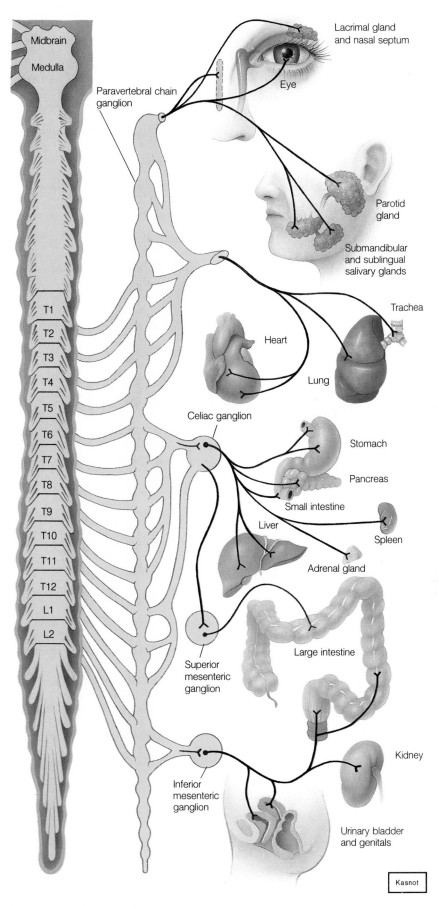

Sympathetic Nervous System
Figure 11.39

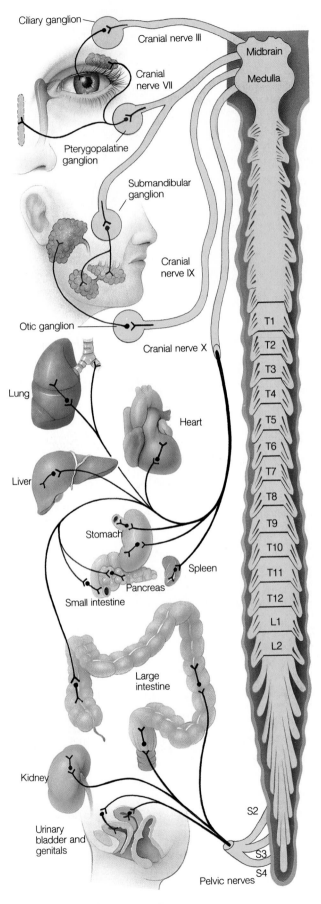

Parasympathetic Nervous System
Figure 11.40

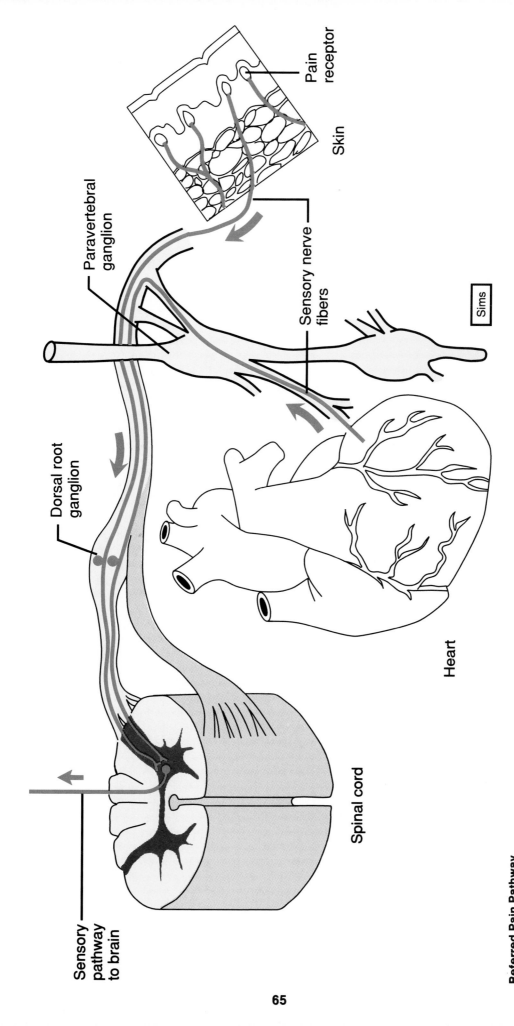

Pain receptor

Skin

Paravertebral ganglion

Sensory nerve fibers

Sims

Dorsal root ganglion

Heart

Sensory pathway to brain

Spinal cord

Referred Pain Pathway
Figure 12.4

65

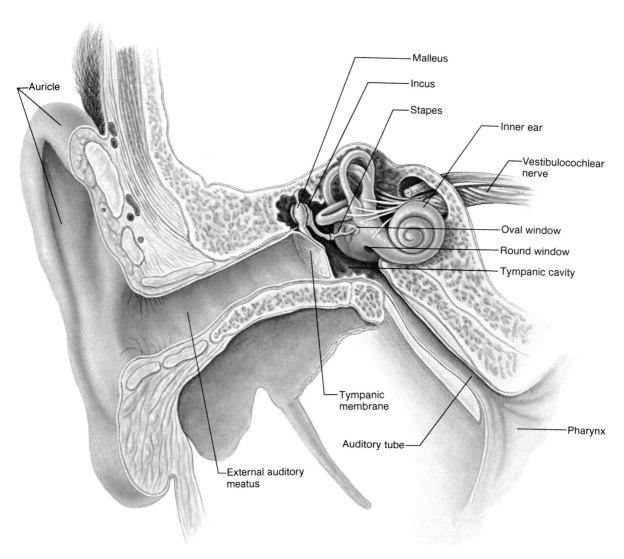

Auricle

Malleus

Incus

Stapes

Inner ear

Vestibulocochlear nerve

Oval window

Round window

Tympanic cavity

Tympanic membrane

Pharynx

Auditory tube

External auditory meatus

Major Parts of the Ear
Figure 12.11

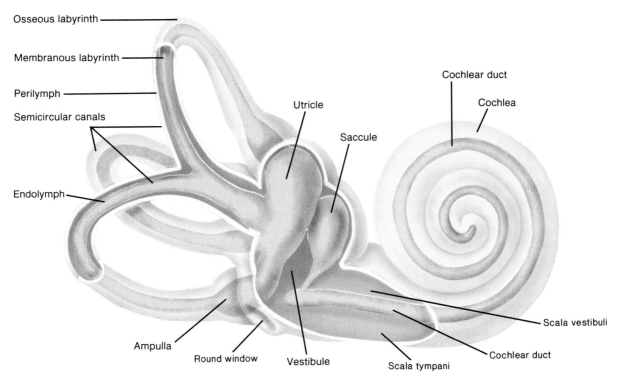

Osseous labyrinth

Membranous labyrinth

Perilymph

Semicircular canals

Endolymph

Utricle

Saccule

Cochlear duct

Cochlea

Scala vestibuli

Ampulla

Round window

Vestibule

Scala tympani

Cochlear duct

Inner Ear Structure
Figure 12.13

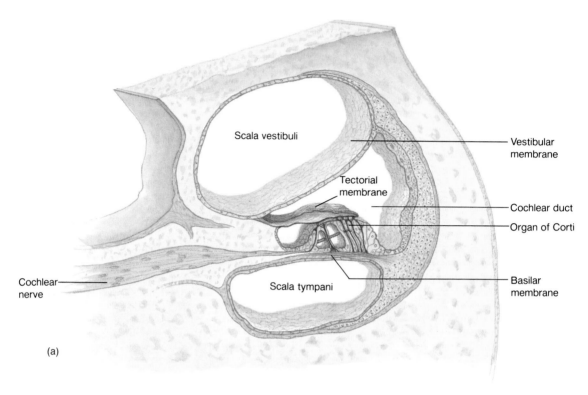

Scala vestibuli

Tectorial membrane

Scala tympani

Cochlear nerve

Vestibular membrane

Cochlear duct

Organ of Corti

Basilar membrane

(a)

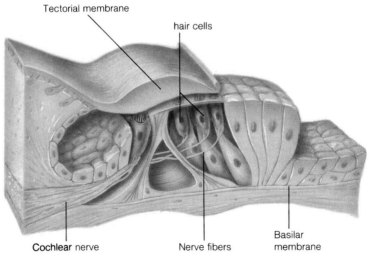

Tectorial membrane

hair cells

(b) Cochlear nerve

Nerve fibers

Basilar membrane

Organ of Corti
Figure 12.15a–b

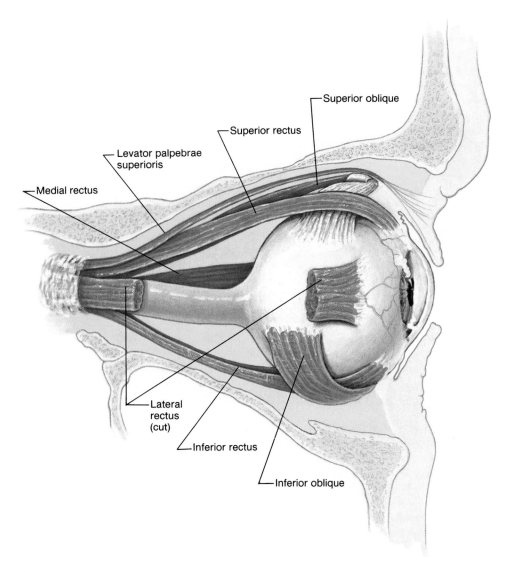

Superior oblique

Superior rectus

Levator palpebrae
superioris

Medial rectus

Lateral
rectus
(cut)

Inferior rectus

Inferior oblique

Extrinsic Muscles of the Eye
Figure 12.26

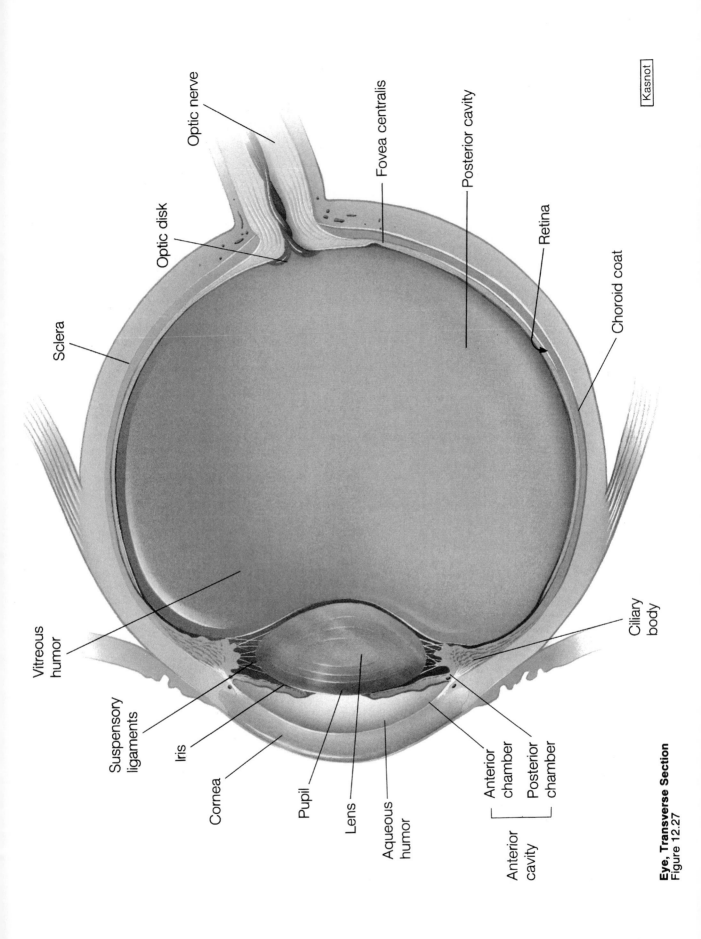

Optic nerve

Optic disk

Fovea centralis

Posterior cavity

Retina

Sclera

Choroid coat

Vitreous humor

Suspensory ligaments

Ciliary body

Iris

Cornea

Pupil

Lens

Aqueous humor

Anterior chamber

Posterior chamber

Anterior cavity

Kasnot

Eye, Transverse Section
Figure 12.27

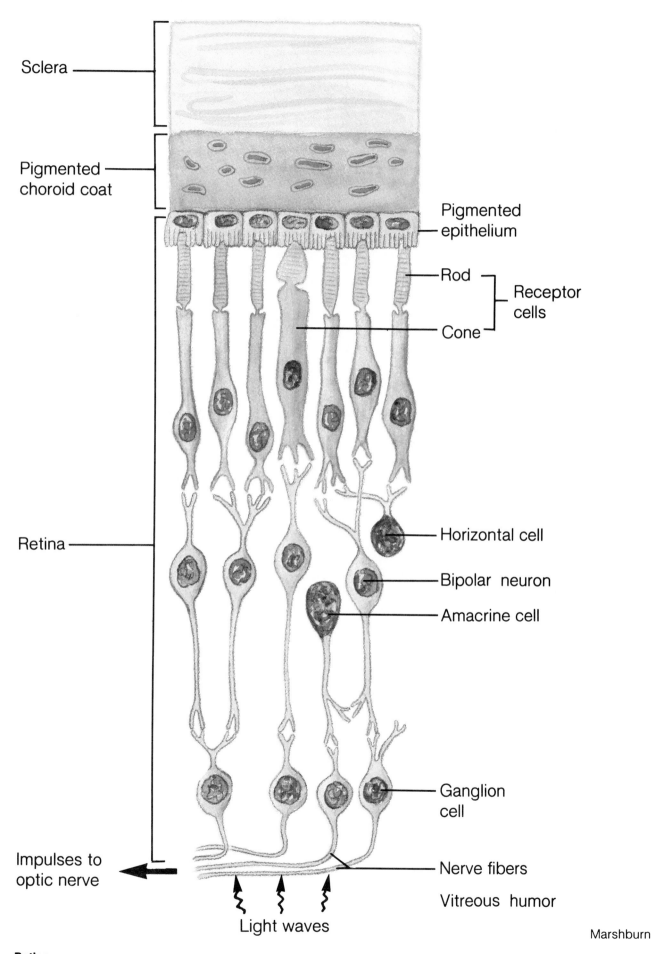

Sclera

Pigmented choroid coat

Pigmented epithelium

Rod

Cone

Receptor cells

Retina

Horizontal cell

Bipolar neuron

Amacrine cell

Ganglion cell

Impulses to optic nerve

Nerve fibers

Vitreous humor

Light waves

Marshburn

Retina
Figure 12.34

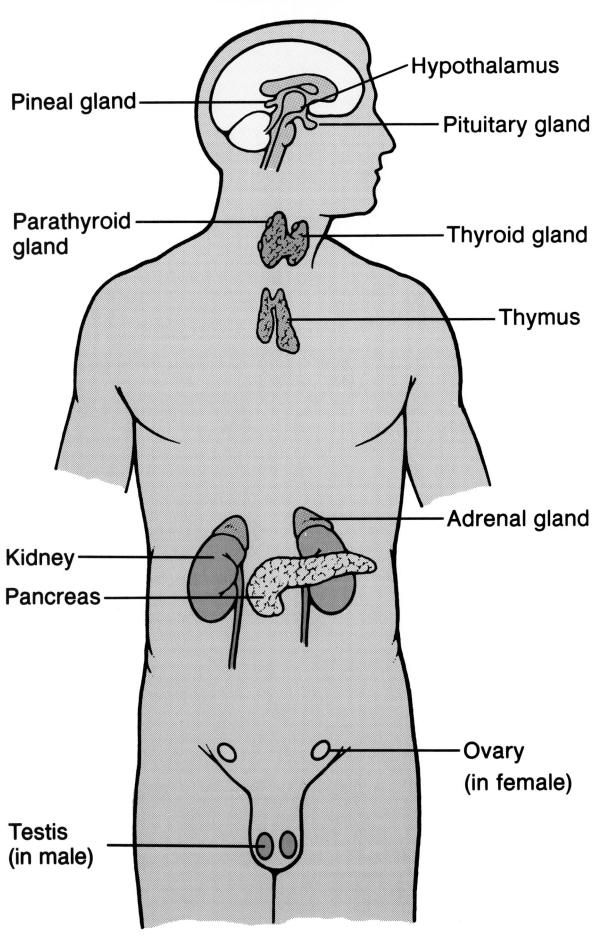

Major Endocrine Glands
Figure 13.2

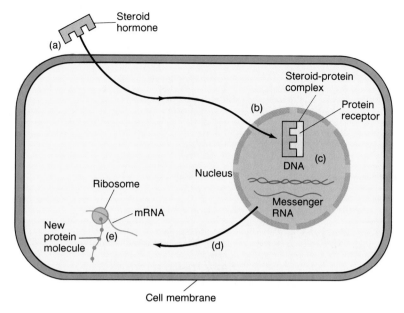

Steroid Hormone Action
Figure 13.4

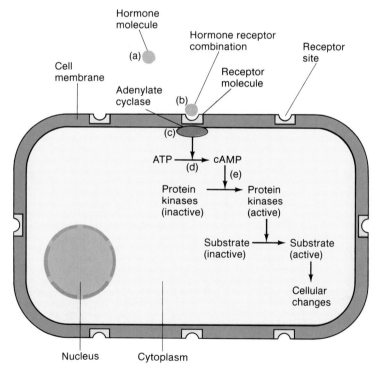

Nonsteroid Hormone Action
Figure 13.6

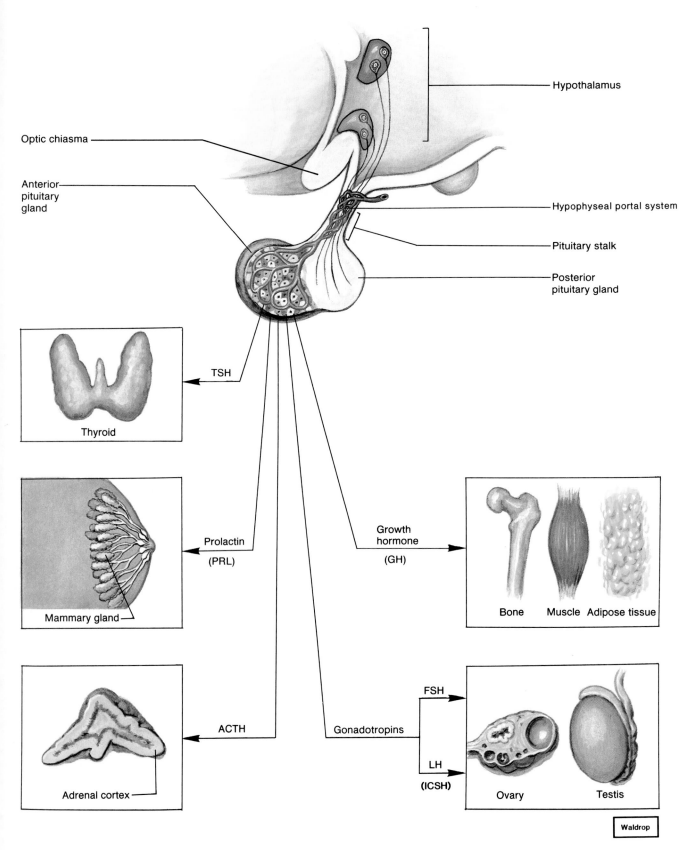

Hypothalamus

Optic chiasma

Anterior pituitary gland

Hypophyseal portal system

Pituitary stalk

Posterior pituitary gland

TSH

Thyroid

Prolactin (PRL)

Mammary gland

Growth hormone (GH)

Bone   Muscle   Adipose tissue

ACTH

Adrenal cortex

Gonadotropins

FSH

LH (ICSH)

Ovary   Testis

Waldrop

Pituitary Hormones and Their Target Organs
Figure 13.13

74

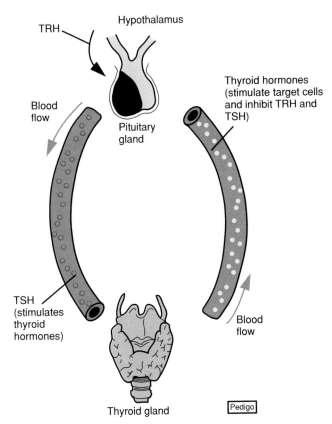

Negative Feedback System
Figure 13.15

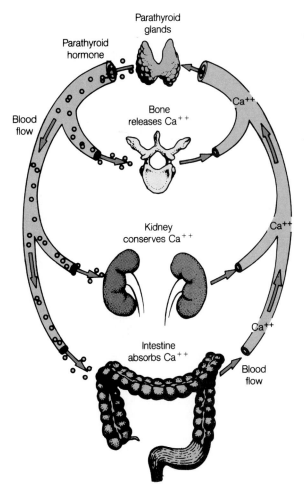

Action of PTH
Figure 13.26

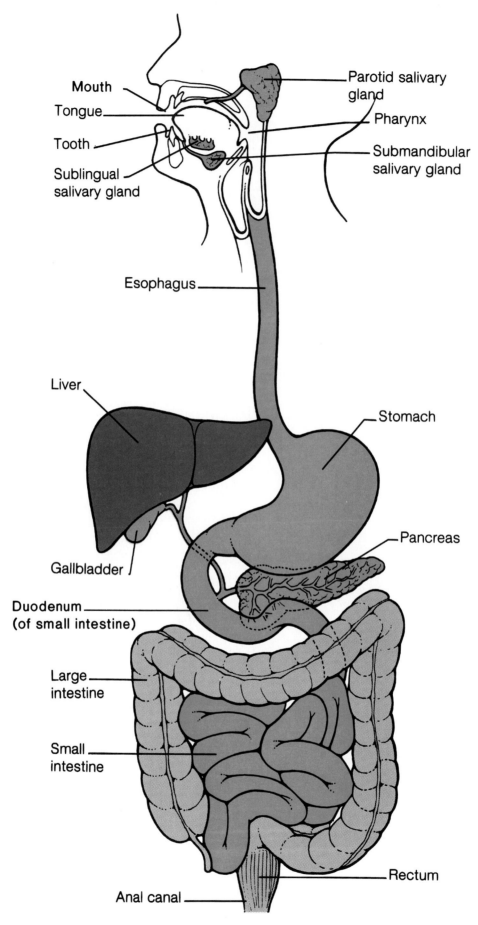

Mouth

Tongue

Tooth

Sublingual
salivary gland

Parotid salivary
gland

Pharynx

Submandibular
salivary gland

Esophagus

Liver

Stomach

Gallbladder

Pancreas

Duodenum
(of small intestine)

Large
intestine

Small
intestine

Rectum

Anal canal

Digestive System
Figure 14.1

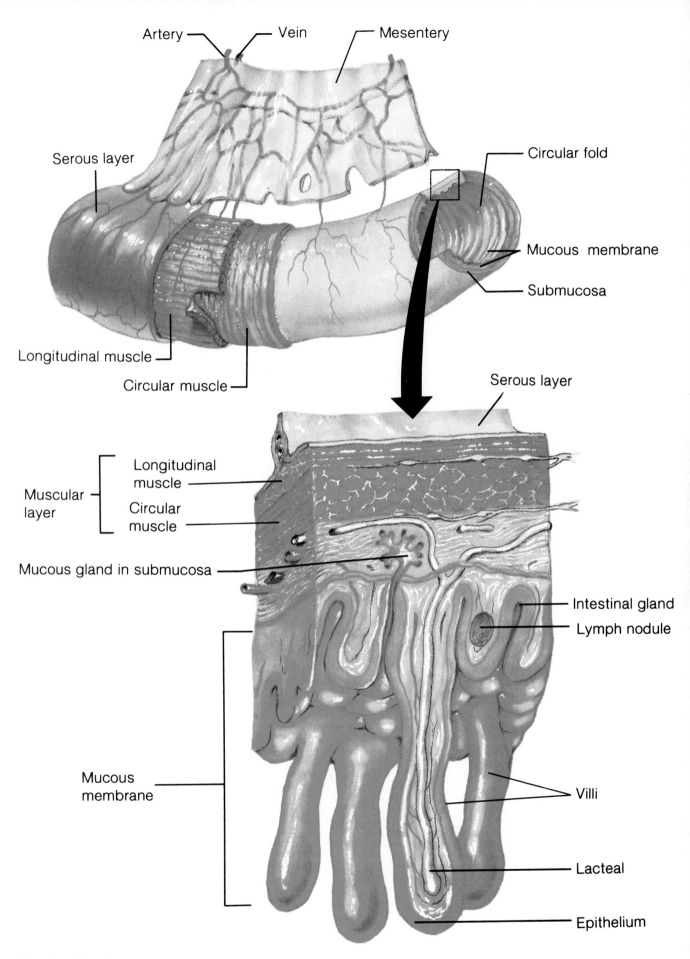

Artery — Vein — Mesentery

Serous layer

Circular fold

Mucous membrane

Submucosa

Longitudinal muscle

Circular muscle

Serous layer

Muscular layer
- Longitudinal muscle
- Circular muscle

Mucous gland in submucosa

Intestinal gland

Lymph nodule

Mucous membrane

Villi

Lacteal

Epithelium

Alimentary Canal
Figure 14.3

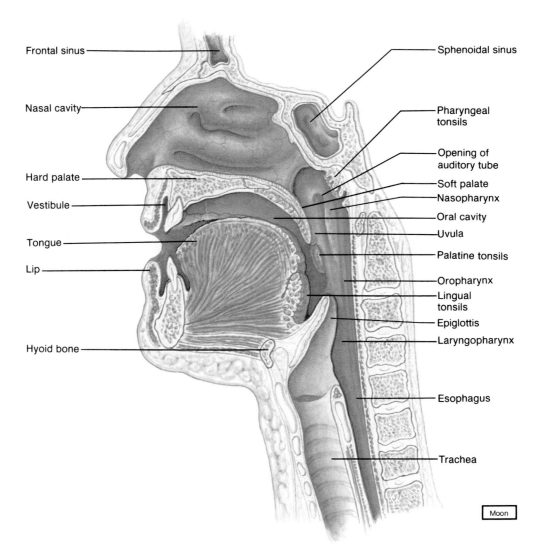

Frontal sinus

Nasal cavity

Hard palate

Vestibule

Tongue

Lip

Hyoid bone

Sphenoidal sinus

Pharyngeal tonsils

Opening of auditory tube

Soft palate

Nasopharynx

Oral cavity

Uvula

Palatine tonsils

Oropharynx

Lingual tonsils

Epiglottis

Laryngopharynx

Esophagus

Trachea

Moon

Mouth and Pharynx
Figure 14.7

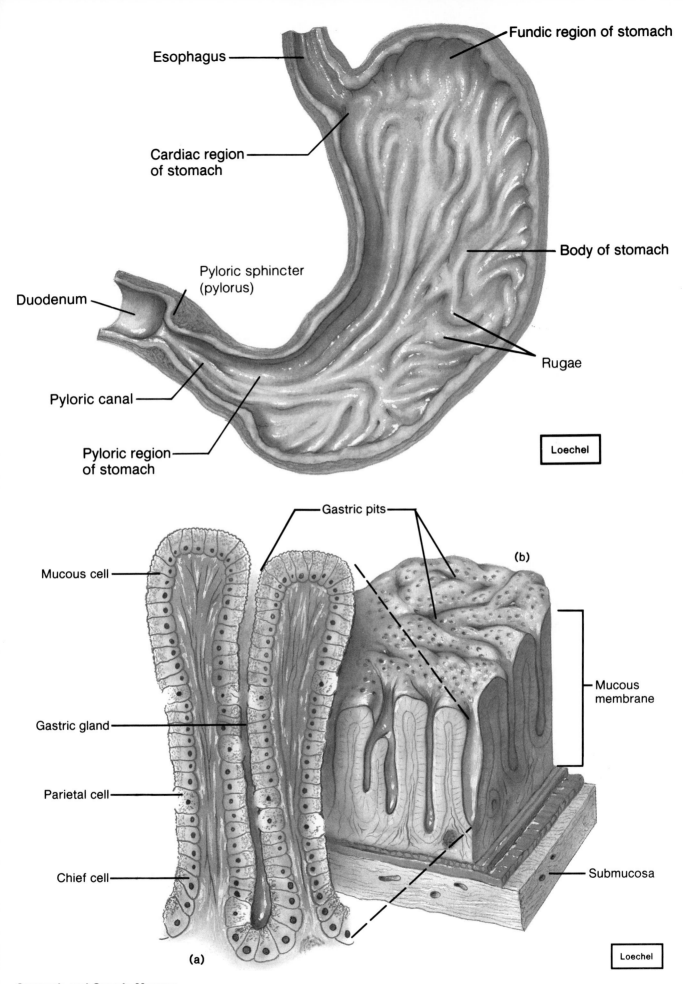

Esophagus

Fundic region of stomach

Cardiac region
of stomach

Body of stomach

Pyloric sphincter
(pylorus)

Duodenum

Rugae

Pyloric canal

Pyloric region
of stomach

Loechel

Gastric pits

(b)

Mucous cell

Gastric gland

Mucous
membrane

Parietal cell

Chief cell

Submucosa

(a)

Loechel

Stomach and Gastric Mucosa
Figures 14.18 and 14.20

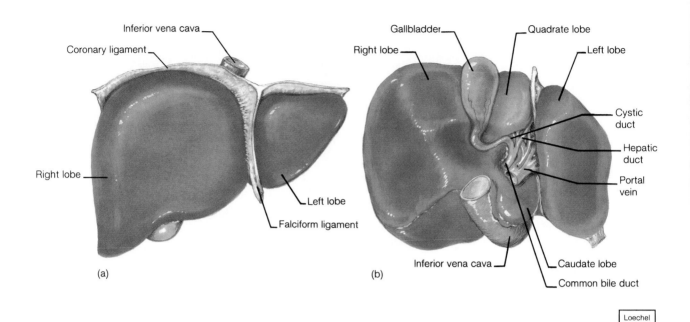

(a)

(b)

Loechel

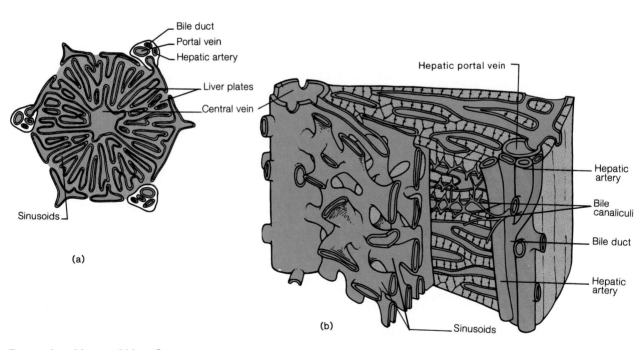

(a)

(b)

External and Internal Liver Structures
Figures 14.30 and 14.31

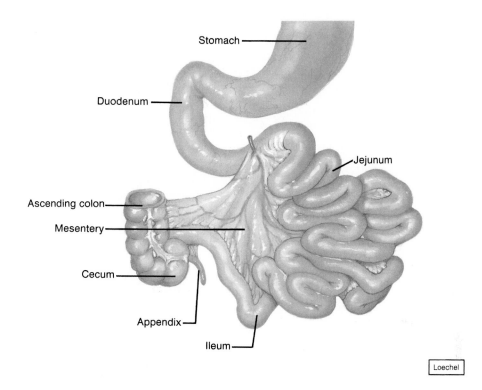

Stomach

Duodenum

Jejunum

Ascending colon

Mesentery

Cecum

Appendix

Ileum

Loechel

Small Intestine
Figure 14.36

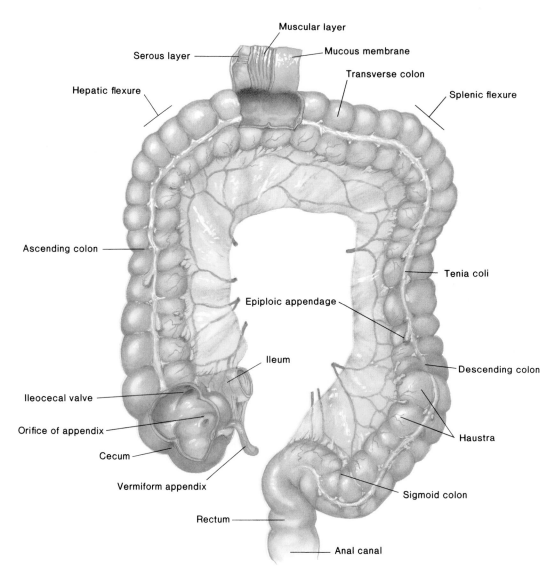

Muscular layer

Serous layer

Mucous membrane

Hepatic flexure

Transverse colon

Splenic flexure

Ascending colon

Tenia coli

Epiploic appendage

Ileum

Descending colon

Ileocecal valve

Orifice of appendix

Haustra

Cecum

Vermiform appendix

Sigmoid colon

Rectum

Anal canal

Large Intestine
Figure 14.48

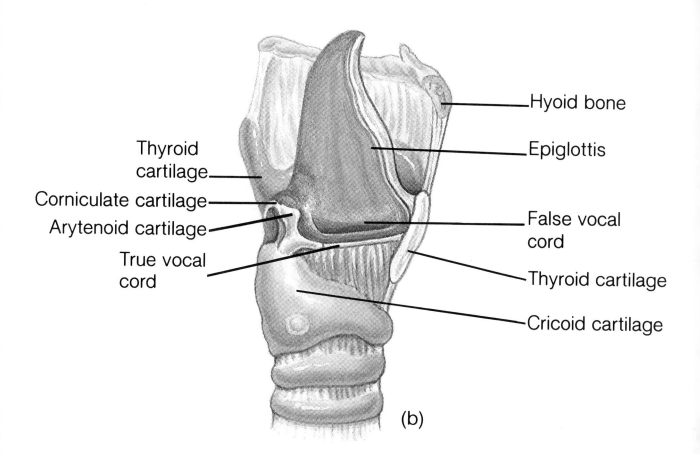

Thyroid cartilage

Corniculate cartilage

Arytenoid cartilage

True vocal cord

Hyoid bone

Epiglottis

False vocal cord

Thyroid cartilage

Cricoid cartilage

(b)

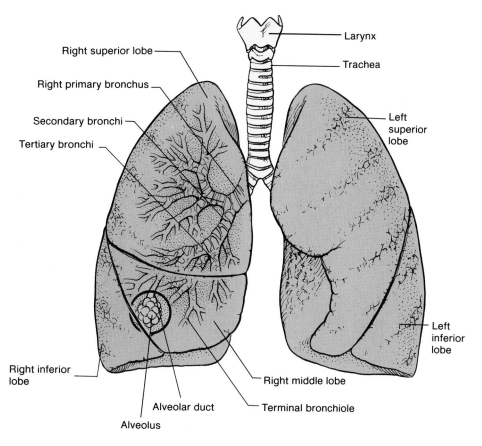

Right superior lobe

Right primary bronchus

Secondary bronchi

Tertiary bronchi

Larynx

Trachea

Left superior lobe

Left inferior lobe

Right inferior lobe

Right middle lobe

Alveolar duct

Terminal bronchiole

Alveolus

Larynx and Bronchial Tree
Figures 16.6b and 16.12

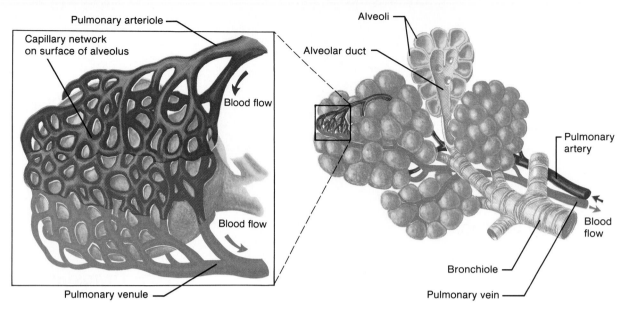

Respiratory Tubes and Alveoli
Figure 16.14

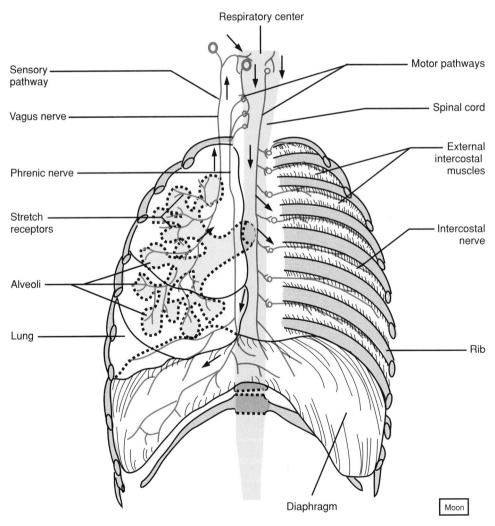

Inflation Reflex
Figure 16.32

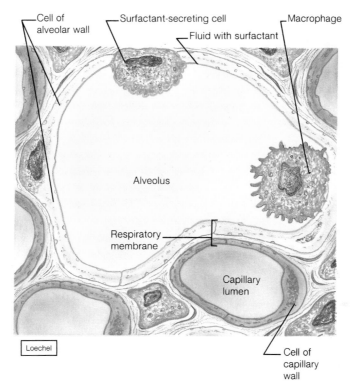

Cell of alveolar wall

Surfactant-secreting cell

Macrophage

Fluid with surfactant

Alveolus

Respiratory membrane

Capillary lumen

Loechel

Cell of capillary wall

Respiratory Membrane
Figure 16.34

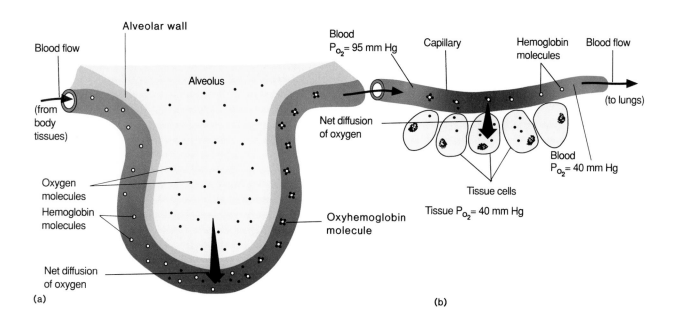

(a)

Blood flow

Alveolar wall

Blood P_{O_2} = 95 mm Hg

Capillary

Hemoglobin molecules

Blood flow

Alveolus

(from body tissues)

(to lungs)

Oxygen molecules

Hemoglobin molecules

Net diffusion of oxygen

Oxyhemoglobin molecule

Net diffusion of oxygen

Net diffusion of oxygen

Blood P_{O_2} = 40 mm Hg

Tissue cells

Tissue P_{O_2} = 40 mm Hg

(b)

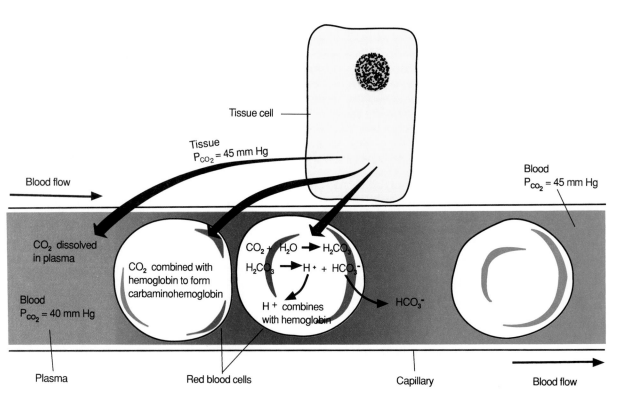

Tissue cell

Tissue P_{CO_2} = 45 mm Hg

Blood P_{CO_2} = 45 mm Hg

Blood flow

CO_2 dissolved in plasma

CO_2 combined with hemoglobin to form carbaminohemoglobin

$CO_2 + H_2O \rightarrow H_2CO_3$

$H_2CO_3 \rightarrow H^+ + HCO_3^-$

HCO_3^-

Blood P_{CO_2} = 40 mm Hg

H^+ combines with hemoglobin

Plasma

Red blood cells

Capillary

Blood flow

Oxygen and Carbon Dioxide Transport
Figures 16.38 and 16.42

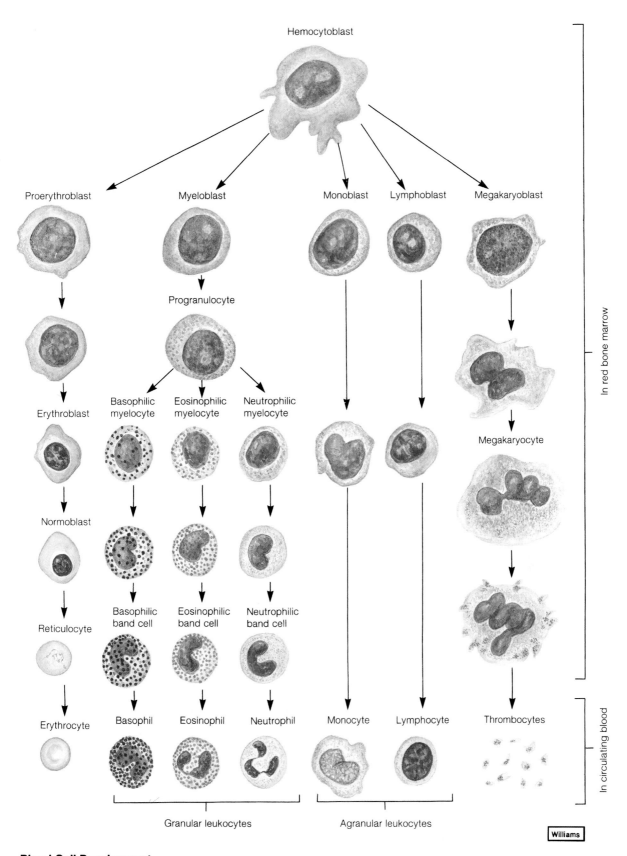

Hemocytoblast

Proerythroblast   Myeloblast   Monoblast   Lymphoblast   Megakaryoblast

Progranulocyte

Erythroblast

Basophilic   Eosinophilic   Neutrophilic
myelocyte   myelocyte   myelocyte

Normoblast

Megakaryocyte

Reticulocyte

Basophilic   Eosinophilic   Neutrophilic
band cell   band cell   band cell

Erythrocyte   Basophil   Eosinophil   Neutrophil   Monocyte   Lymphocyte   Thrombocytes

Granular leukocytes   Agranular leukocytes

In red bone marrow

In circulating blood

Williams

Blood Cell Development
Figure 17.6

87

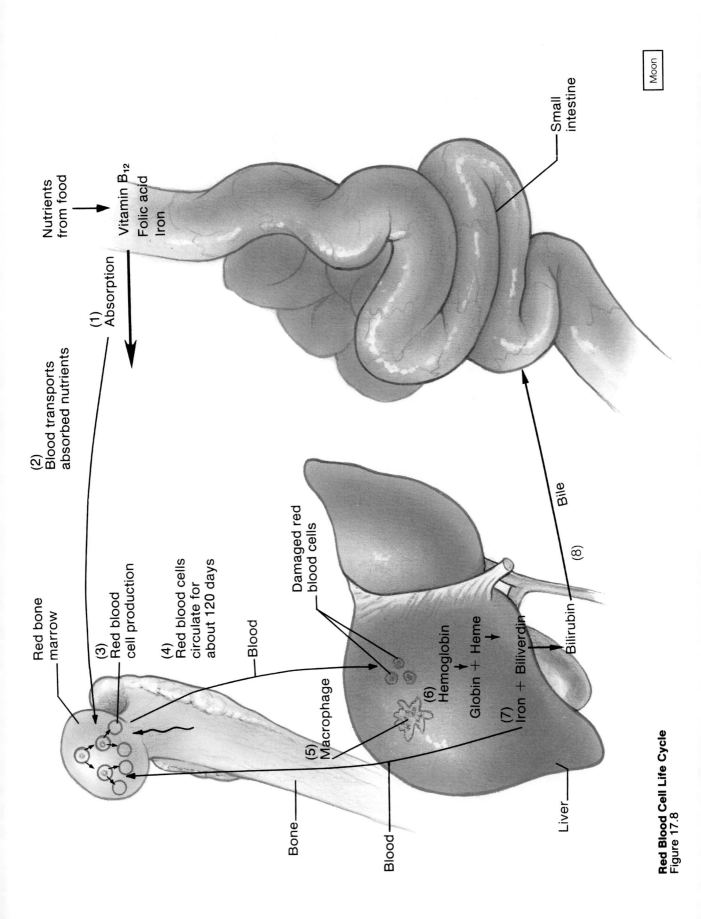

Nutrients from food

Vitamin B$_{12}$
Folic acid
Iron

Small intestine

(1) Absorption

(2) Blood transports absorbed nutrients

Red bone marrow

(3) Red blood cell production

(4) Red blood cells circulate for about 120 days

Blood

Bone

Blood

Damaged red blood cells

(5) Macrophage

(6) Hemoglobin

Globin + Heme

(7) Iron + Biliverdin

Bilirubin

Bile

(8)

Liver

Moon

Red Blood Cell Life Cycle
Figure 17.8

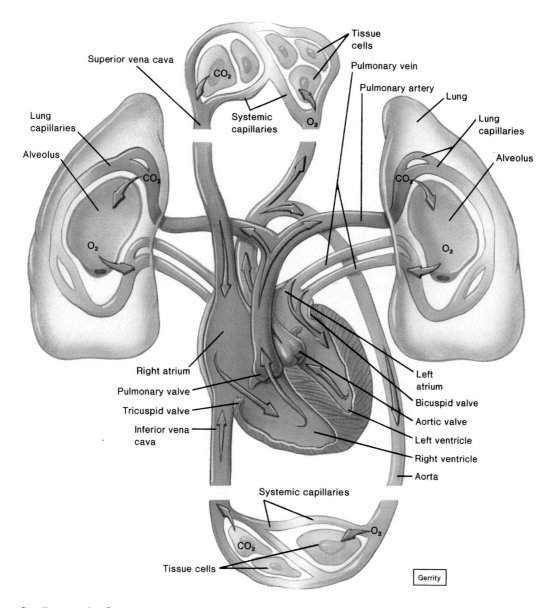

Cardiovascular System
Figure 18.10

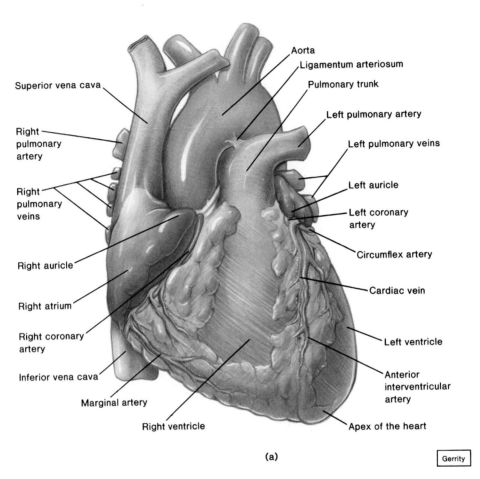

Superior vena cava

Right
pulmonary
artery

Right
pulmonary
veins

Right auricle

Right atrium

Right coronary
artery

Inferior vena cava

Marginal artery

Right ventricle

Aorta

Ligamentum arteriosum

Pulmonary trunk

Left pulmonary artery

Left pulmonary veins

Left auricle

Left coronary
artery

Circumflex artery

Cardiac vein

Left ventricle

Anterior
interventricular
artery

Apex of the heart

(a)

Gerrity

Heart and Coronary Circulation, Anterior View
Figure 18.12a

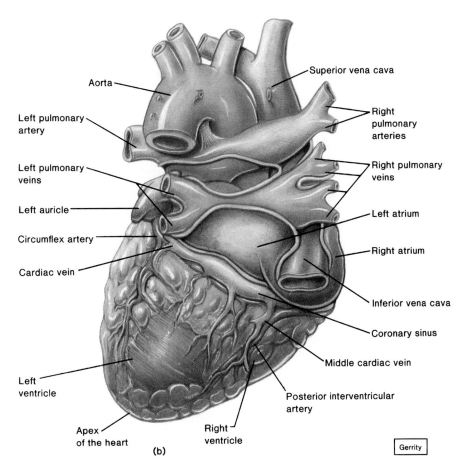

Aorta

Left pulmonary artery

Left pulmonary veins

Left auricle

Circumflex artery

Cardiac vein

Left ventricle

Apex of the heart

Right ventricle

(b)

Superior vena cava

Right pulmonary arteries

Right pulmonary veins

Left atrium

Right atrium

Inferior vena cava

Coronary sinus

Middle cardiac vein

Posterior interventricular artery

Gerrity

Heart and Coronary Circulation, Posterior View
Figure 18.12b

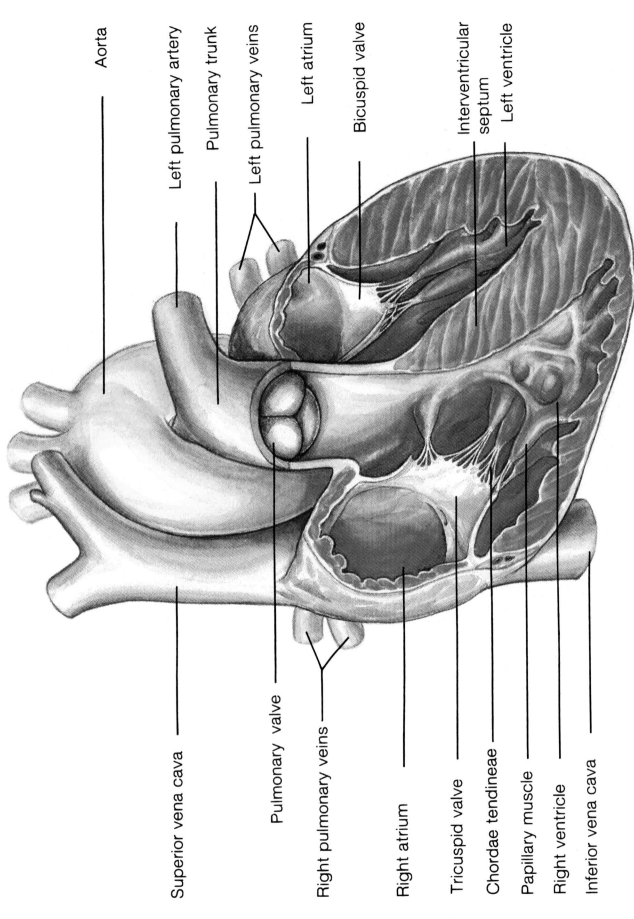

Aorta

Left pulmonary artery

Pulmonary trunk

Left pulmonary veins

Left atrium

Bicuspid valve

Interventricular septum

Left ventricle

Superior vena cava

Pulmonary valve

Right pulmonary veins

Right atrium

Tricuspid valve

Chordae tendineae

Papillary muscle

Right ventricle

Inferior vena cava

Heart, Frontal Section I
Figure 18.6a

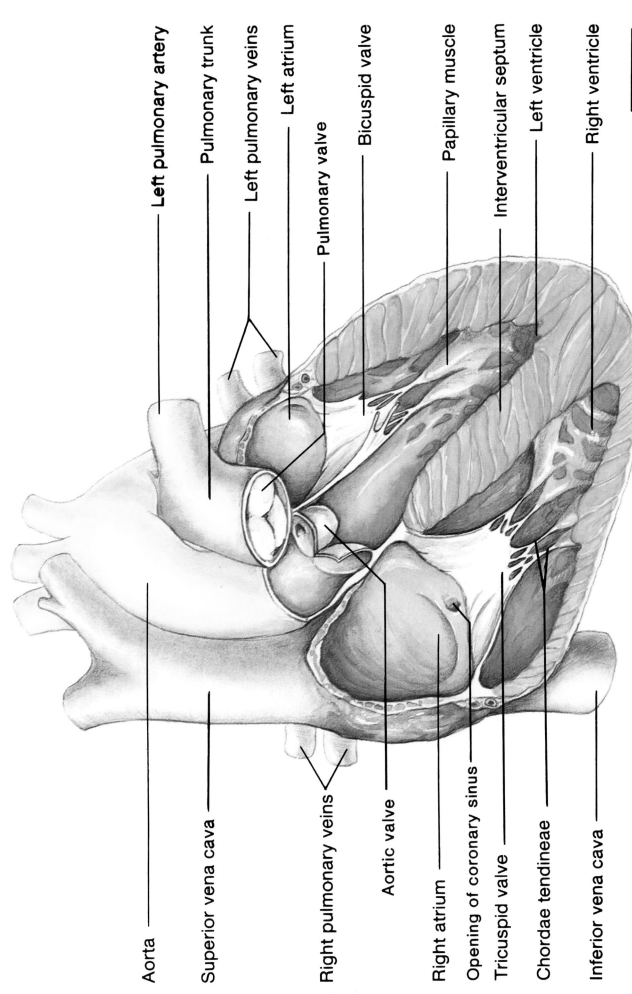

Left pulmonary artery

Pulmonary trunk

Left pulmonary veins

Left atrium

Pulmonary valve

Bicuspid valve

Papillary muscle

Interventricular septum

Left ventricle

Right ventricle

Marshburn

Aorta

Superior vena cava

Right pulmonary veins

Aortic valve

Right atrium

Opening of coronary sinus

Tricuspid valve

Chordae tendineae

Inferior vena cava

Heart, Frontal Section II
Figure 18.6b

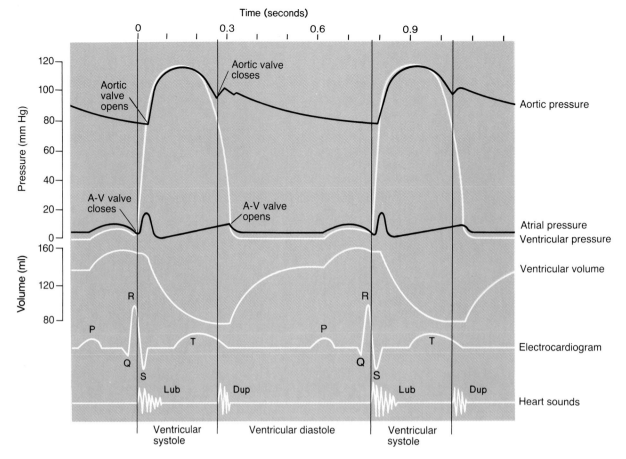

Events of the Cardiac Cycle
Figure 18.17

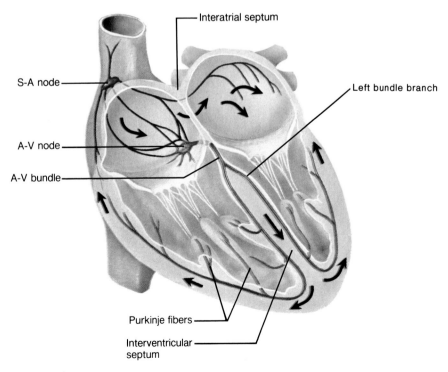

Cardiac Conduction System
Figure 18.19

94

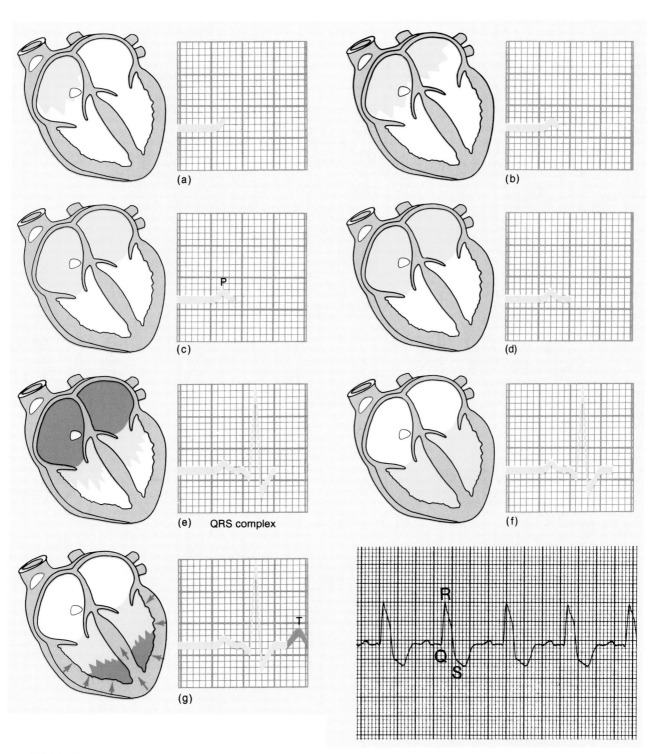

(a)

(b)

(c) P

(d)

(e) QRS complex

(f)

(g) T

R
Q
S

ECG Pattern
Figure 18.25

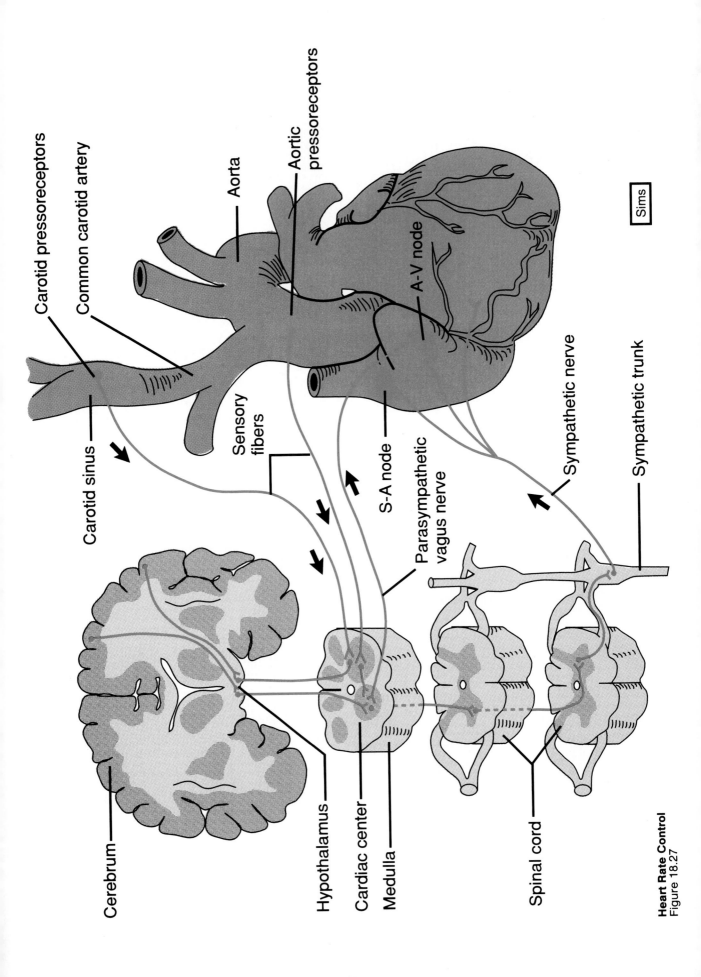

Carotid pressoreceptors
Common carotid artery
Aorta
Aortic pressoreceptors
Carotid sinus
Sensory fibers
S-A node
A-V node
Parasympathetic vagus nerve
Sympathetic nerve
Sympathetic trunk
Cerebrum
Hypothalamus
Cardiac center
Medulla
Spinal cord

Sims

Heart Rate Control
Figure 18.27

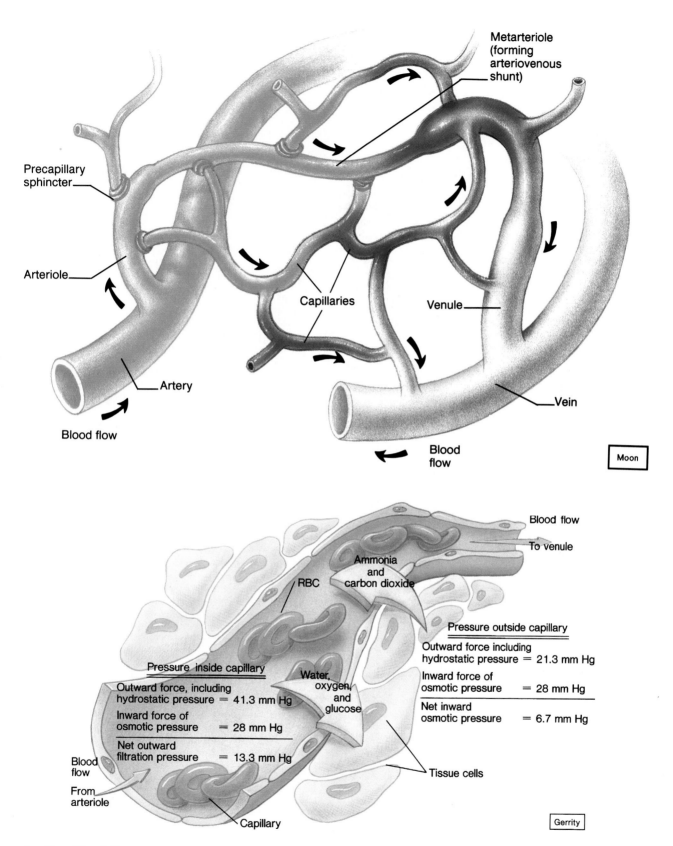

Capillary Circulation
Figures 18.34 and 18.37

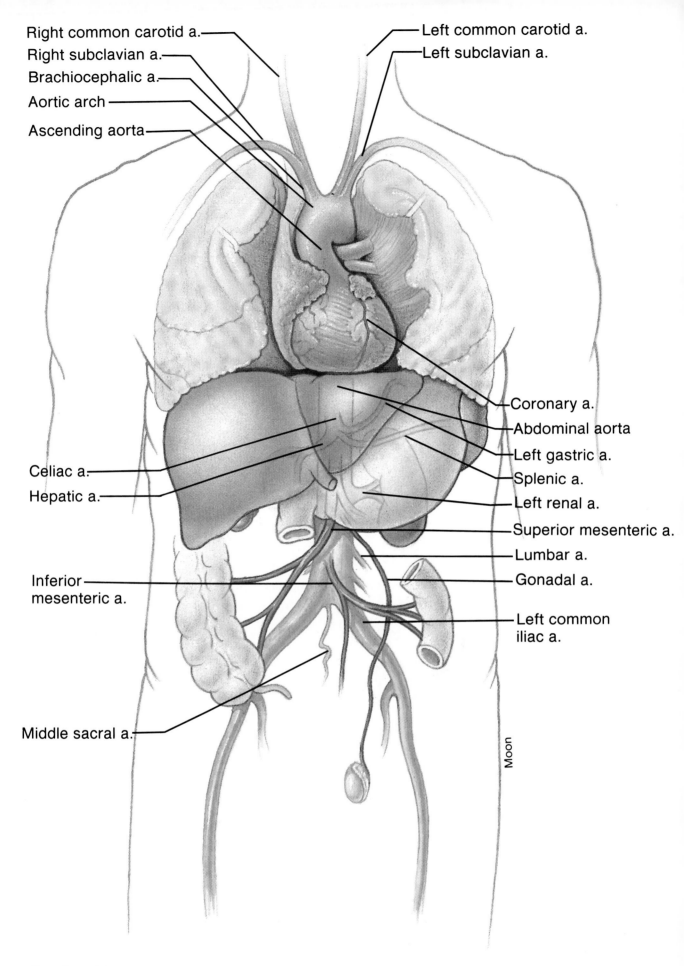

Right common carotid a.
Right subclavian a.
Brachiocephalic a.
Aortic arch
Ascending aorta

Left common carotid a.
Left subclavian a.

Coronary a.
Abdominal aorta
Left gastric a.
Splenic a.
Left renal a.
Superior mesenteric a.
Lumbar a.
Gonadal a.
Left common iliac a.

Celiac a.
Hepatic a.

Inferior mesenteric a.

Middle sacral a.

Moon

Branches of the Aorta
Figure 18.56

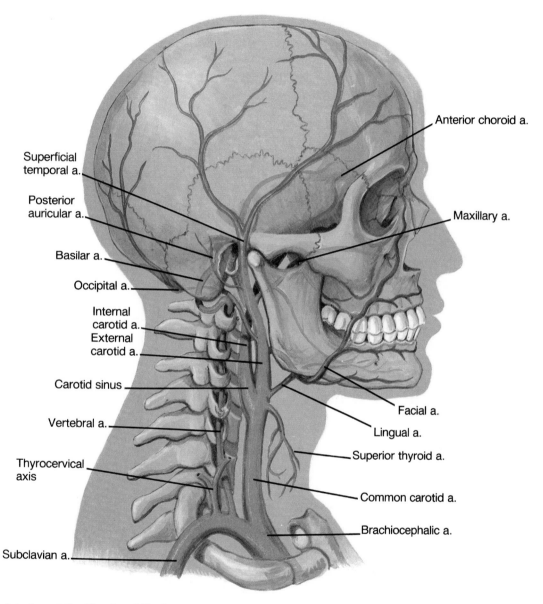

Anterior choroid a.

Superficial
temporal a.

Posterior
auricular a.

Maxillary a.

Basilar a.

Occipital a.

Internal
carotid a.

External
carotid a.

Carotid sinus

Facial a.

Vertebral a.

Lingual a.

Superior thyroid a.

Thyrocervical
axis

Common carotid a.

Brachiocephalic a.

Subclavian a.

Arteries of the Head and Neck
Figure 18.58

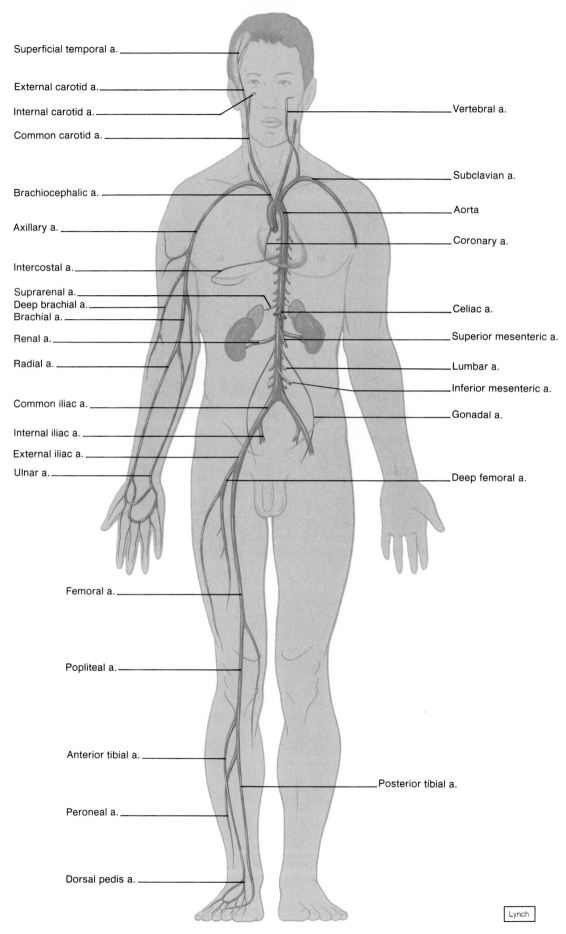

Superficial temporal a.

External carotid a.

Internal carotid a.

Common carotid a.

Brachiocephalic a.

Axillary a.

Intercostal a.

Suprarenal a.
Deep brachial a.
Brachial a.

Renal a.

Radial a.

Common iliac a.

Internal iliac a.

External iliac a.

Ulnar a.

Femoral a.

Popliteal a.

Anterior tibial a.

Peroneal a.

Dorsal pedis a.

Vertebral a.

Subclavian a.

Aorta

Coronary a.

Celiac a.

Superior mesenteric a.

Lumbar a.

Inferior mesenteric a.

Gonadal a.

Deep femoral a.

Posterior tibial a.

Lynch

Arterial System
Figure 18.65

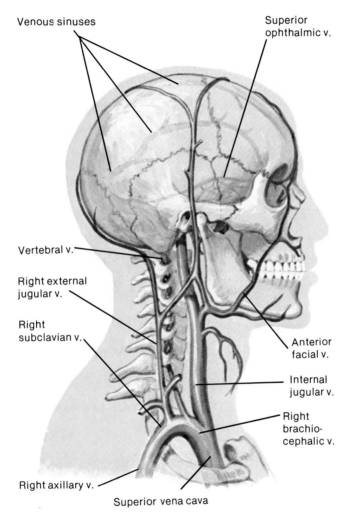

Venous sinuses

Superior
ophthalmic v.

Vertebral v.

Right external
jugular v.

Right
subclavian v.

Anterior
facial v.

Internal
jugular v.

Right
brachio-
cephalic v.

Right axillary v.

Superior vena cava

Veins of the Head and Neck
Figure 18.66

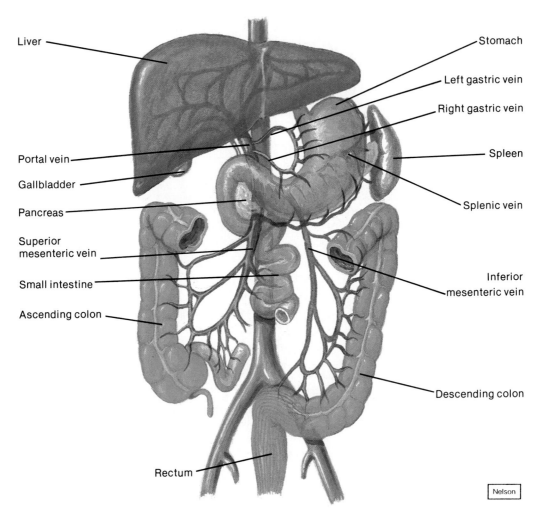

Liver

Portal vein

Gallbladder

Pancreas

Superior
mesenteric vein

Small intestine

Ascending colon

Rectum

Stomach

Left gastric vein

Right gastric vein

Spleen

Splenic vein

Inferior
mesenteric vein

Descending colon

Nelson

Veins of the Abdominal Viscera
Figure 18.69

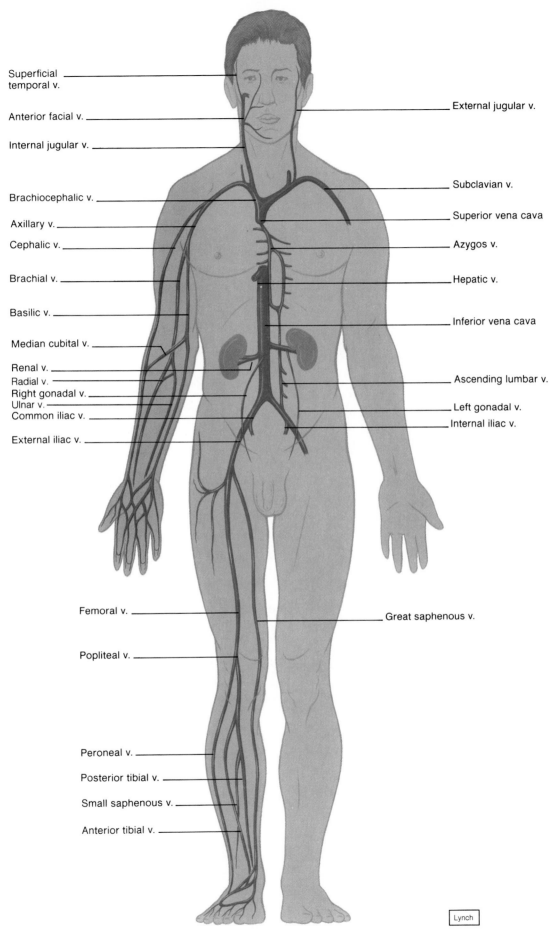

Superficial temporal v.

Anterior facial v.

Internal jugular v.

Brachiocephalic v.

Axillary v.

Cephalic v.

Brachial v.

Basilic v.

Median cubital v.

Renal v.

Radial v.

Right gonadal v.

Ulnar v.

Common iliac v.

External iliac v.

External jugular v.

Subclavian v.

Superior vena cava

Azygos v.

Hepatic v.

Inferior vena cava

Ascending lumbar v.

Left gonadal v.

Internal iliac v.

Femoral v.

Popliteal v.

Great saphenous v.

Peroneal v.

Posterior tibial v.

Small saphenous v.

Anterior tibial v.

Lynch

Venous System
Figure 18.72

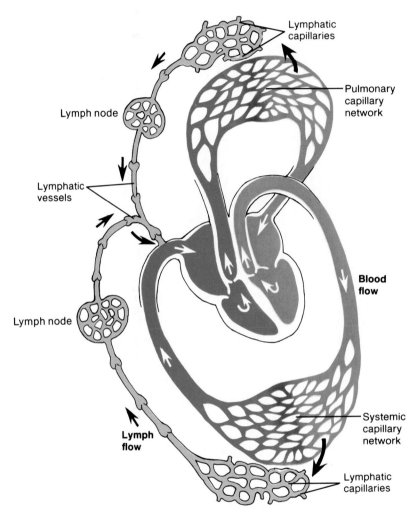

Lymphatic capillaries

Pulmonary capillary network

Lymph node

Lymphatic vessels

Blood flow

Lymph node

Systemic capillary network

Lymph flow

Lymphatic capillaries

Lymphatic System
Figure 19.1

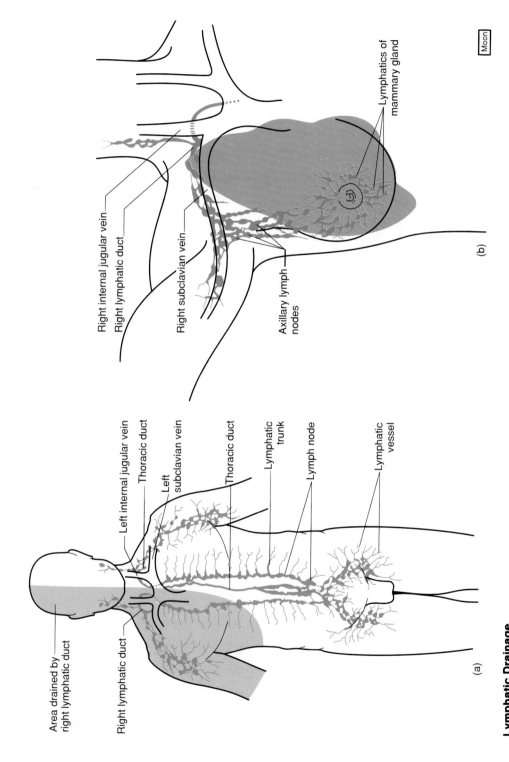

Right internal jugular vein

Right lymphatic duct

Right subclavian vein

Axillary lymph nodes

Lymphatics of mammary gland

Moon

(b)

Left internal jugular vein

Thoracic duct

Left subclavian vein

Thoracic duct

Lymphatic trunk

Lymph node

Lymphatic vessel

Area drained by right lymphatic duct

Right lymphatic duct

(a)

Lymphatic Drainage
Figure 19.6a–b

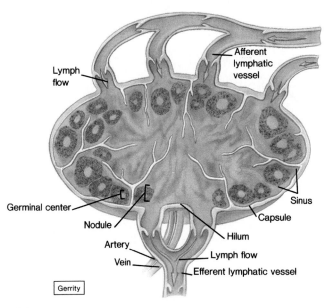

Lymph flow

Afferent lymphatic vessel

Sinus

Capsule

Germinal center

Nodule

Hilum

Artery

Vein

Lymph flow

Efferent lymphatic vessel

Gerrity

(a)

Lymph Node
Figure 19.9a

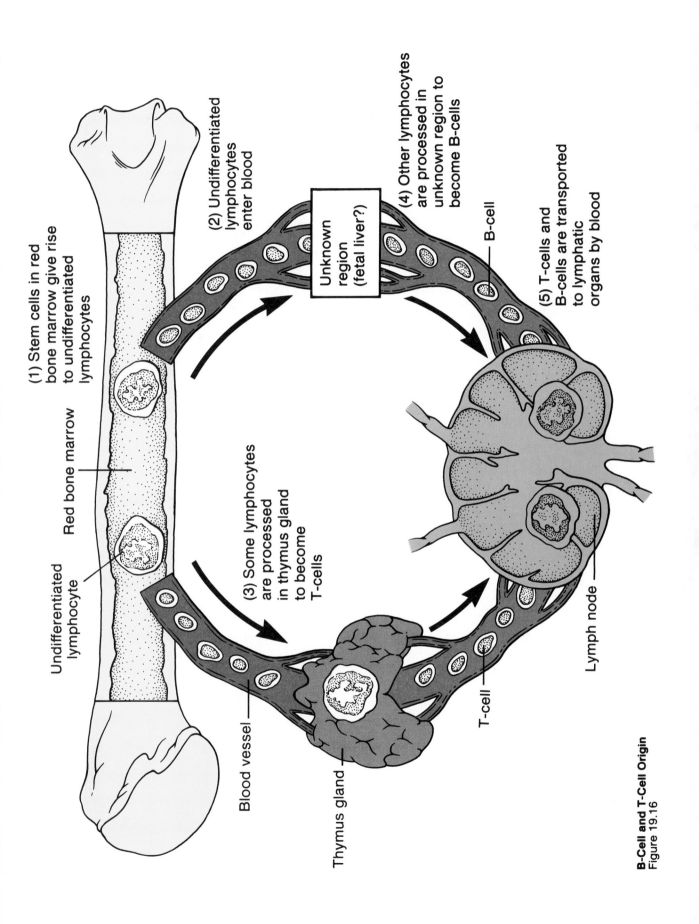

(1) Stem cells in red bone marrow give rise to undifferentiated lymphocytes

(2) Undifferentiated lymphocytes enter blood

(4) Other lymphocytes are processed in unknown region to become B-cells

(5) T-cells and B-cells are transported to lymphatic organs by blood

Red bone marrow

Undifferentiated lymphocyte

Unknown region (fetal liver?)

B-cell

(3) Some lymphocytes are processed in thymus gland to become T-cells

Blood vessel

Thymus gland

T-cell

Lymph node

B-Cell and T-Cell Origin
Figure 19.16

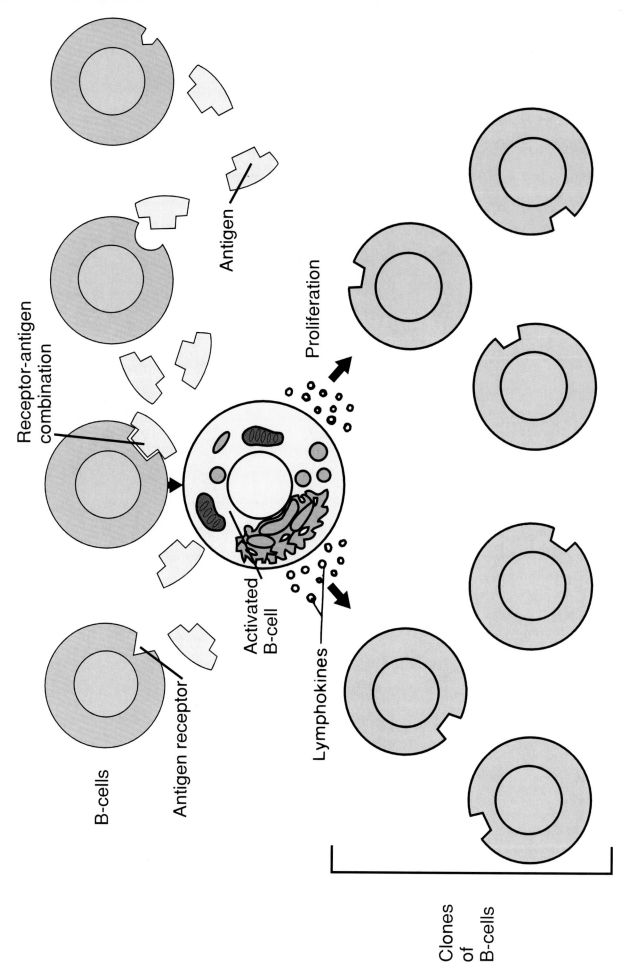

B-cells

Antigen receptor

Antigen

Receptor-antigen combination

Activated B-cell

Lymphokines

Proliferation

Clones
of
B-cells

B-Cell Proliferation
Figure 19.20

108

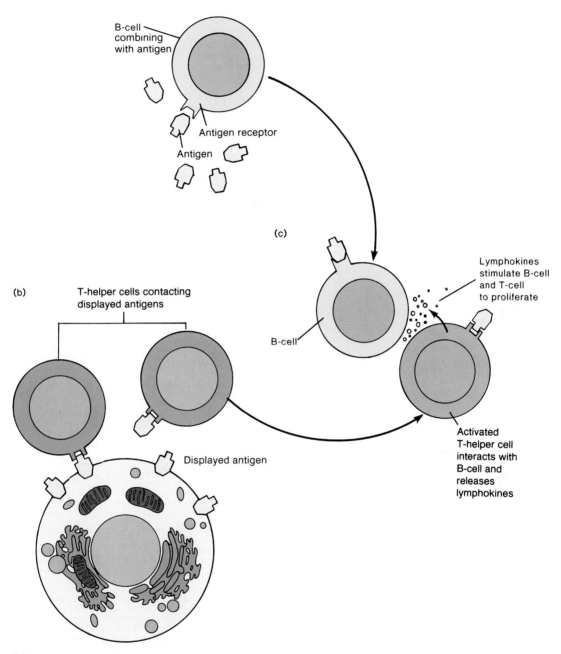

B-cell
combining
with antigen

Antigen receptor

Antigen

(c)

Lymphokines
stimulate B-cell
and T-cell
to proliferate

(b)

T-helper cells contacting
displayed antigens

B-cell

Displayed antigen

Activated
T-helper cell
interacts with
B-cell and
releases
lymphokines

(a)   **Macrophage displaying antigen**

B-Cell and T-Cell Interaction
Figure 19.21

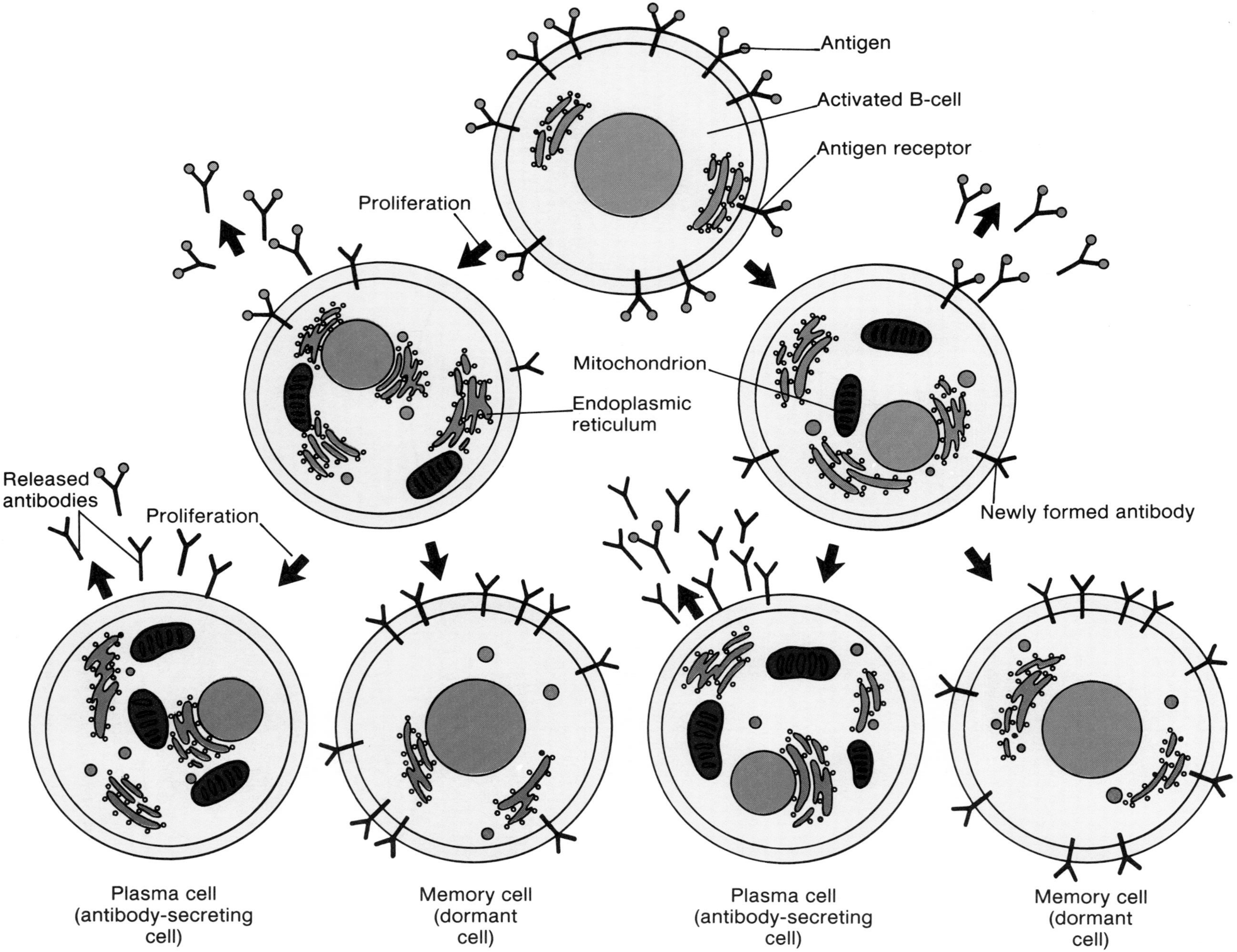

Antibody Formation
Figure 19.22

Antigen

Activated B-cell

Antigen receptor

Newly formed antibody

Memory cell (dormant cell)

Mitochondrion

Endoplasmic reticulum

Proliferation

Plasma cell (antibody-secreting cell)

Memory cell (dormant cell)

Proliferation

Released antibodies

Plasma cell (antibody-secreting cell)

110

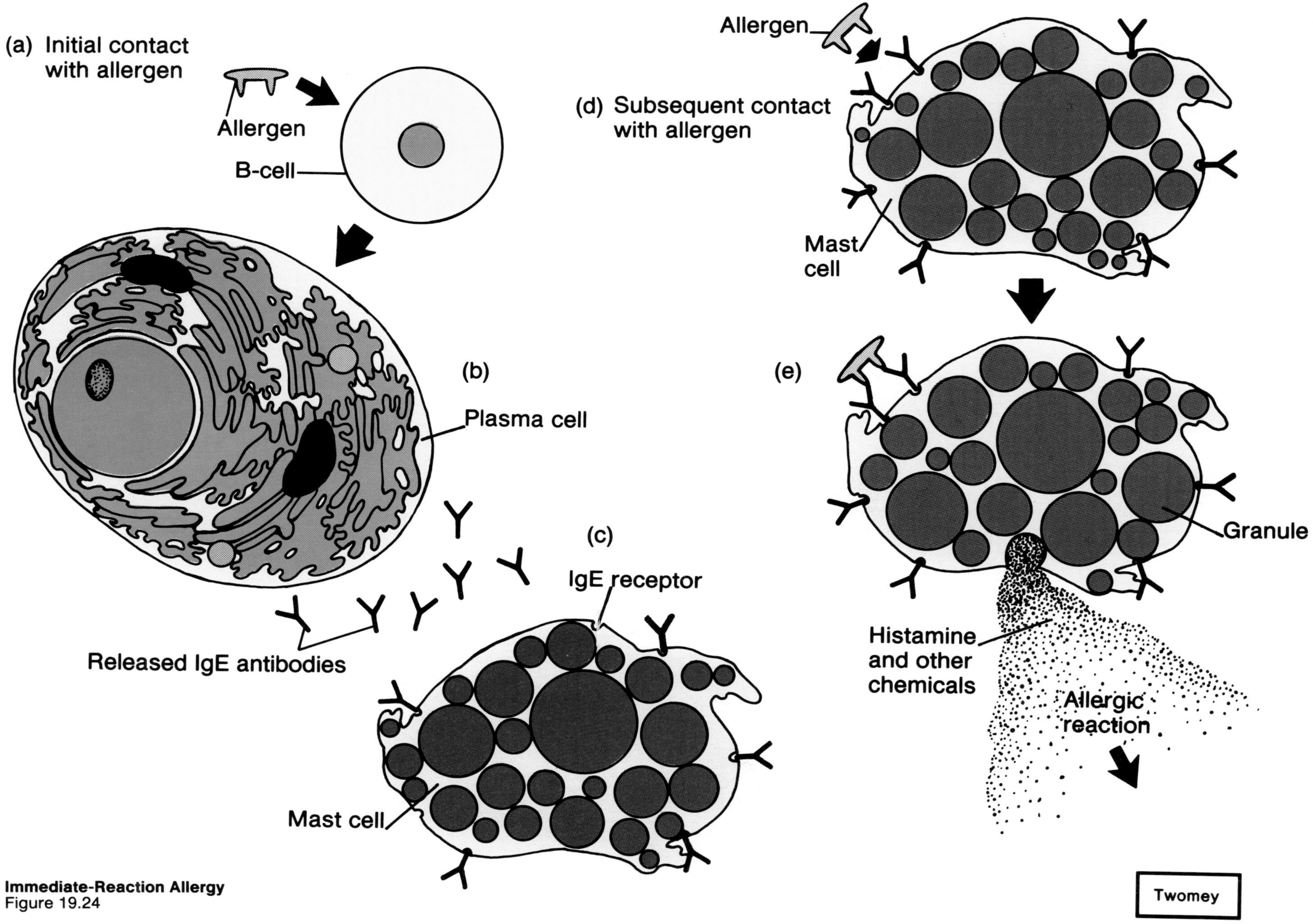

(a) Initial contact with allergen

Allergen

B-cell

(b) Plasma cell

Released IgE antibodies

(c) IgE receptor

Mast cell

(d) Subsequent contact with allergen

Allergen

Mast cell

Granule

(e)

Histamine and other chemicals

Allergic reaction

Twomey

Immediate-Reaction Allergy
Figure 19.24

111

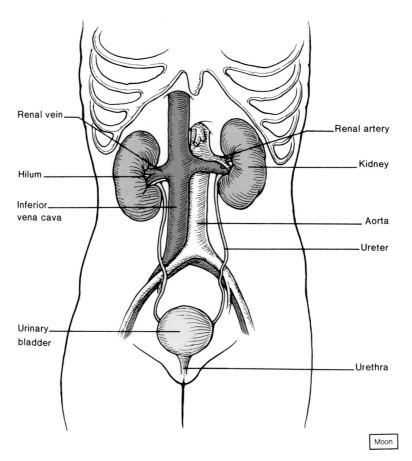

Renal vein

Hilum

Inferior
vena cava

Urinary
bladder

Renal artery

Kidney

Aorta

Ureter

Urethra

Moon

Urinary System
Figure 20.1

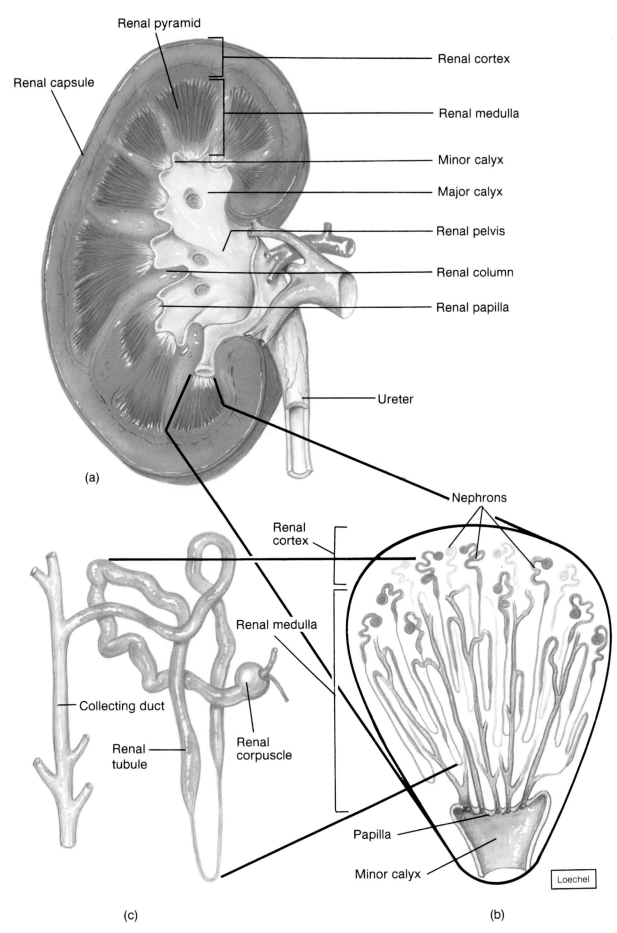

Renal pyramid

Renal cortex

Renal capsule

Renal medulla

Minor calyx

Major calyx

Renal pelvis

Renal column

Renal papilla

Ureter

(a)

Nephrons

Renal
cortex

Renal medulla

Collecting duct

Renal
corpuscle

Renal
tubule

Papilla

Minor calyx

Loechel

(c)

(b)

Kidney Structure
Figure 20.4

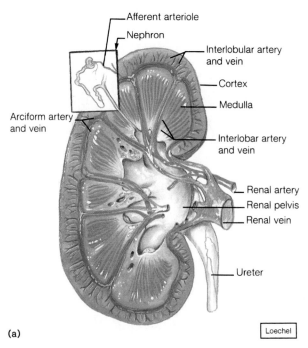

Afferent arteriole

Nephron

Interlobular artery
and vein

Cortex

Medulla

Arciform artery
and vein

Interlobar artery
and vein

Renal artery

Renal pelvis

Renal vein

Ureter

(a)

Loechel

Renal Arteries and Veins
Figure 20.6a

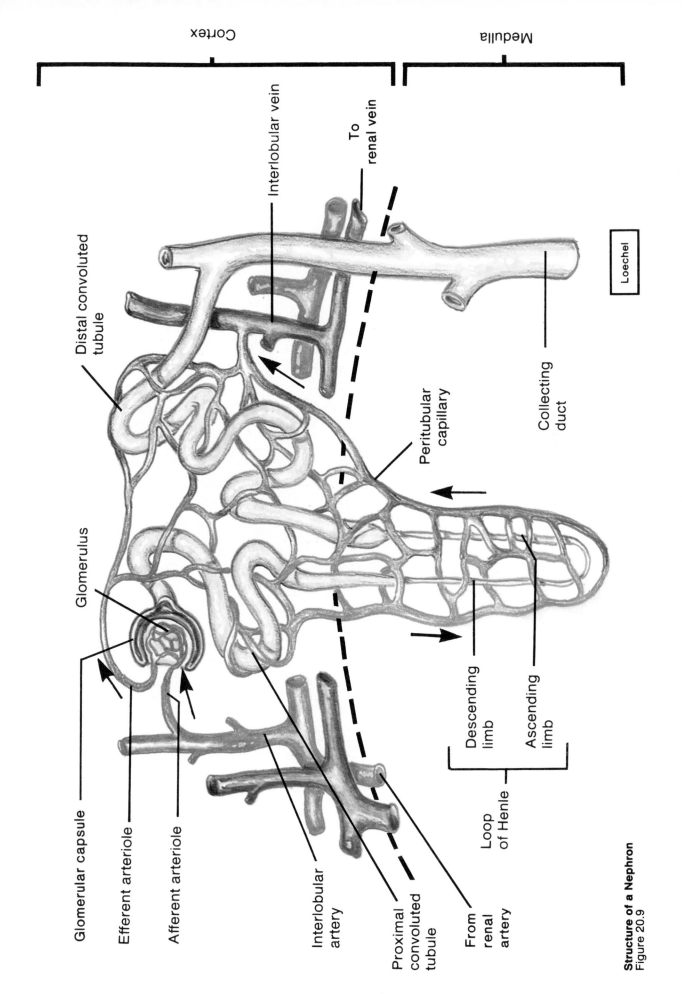

Interlobular vein

To renal vein

Loechel

Distal convoluted tubule

Peritubular capillary

Collecting duct

Glomerulus

Glomerular capsule

Efferent arteriole

Afferent arteriole

Interlobular artery

Proximal convoluted tubule

From renal artery

Descending limb

Ascending limb

Loop of Henle

Structure of a Nephron
Figure 20.9

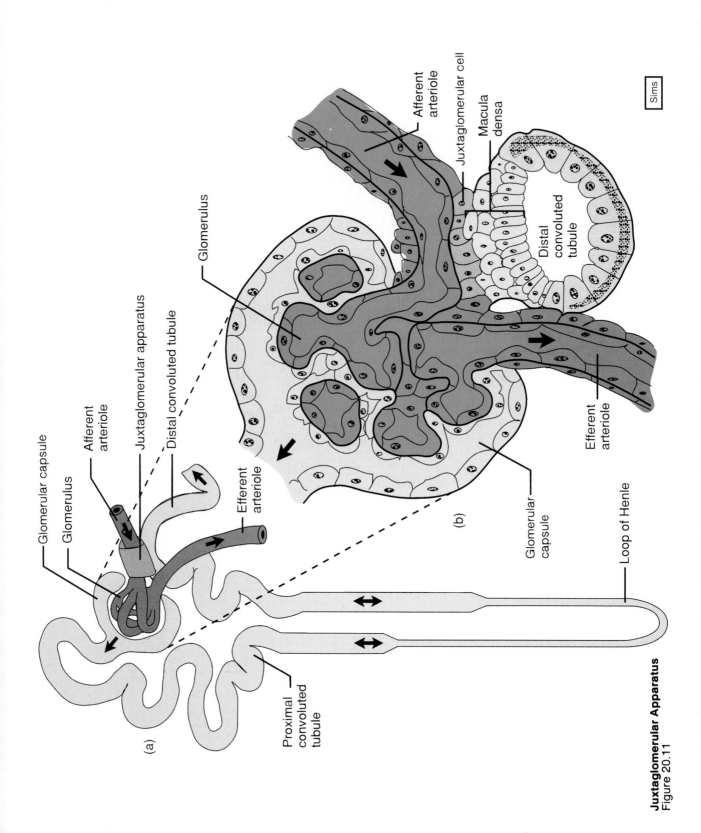

Glomerular capsule

Glomerulus

Afferent
arteriole

Juxtaglomerular apparatus

Distal convoluted tubule

Efferent
arteriole

Proximal
convoluted
tubule

(a)

Glomerulus

Afferent
arteriole

Juxtaglomerular cell

Macula
densa

Distal
convoluted
tubule

Efferent
arteriole

Glomerular
capsule

Loop of Henle

(b)

Sims

Juxtaglomerular Apparatus
Figure 20.11

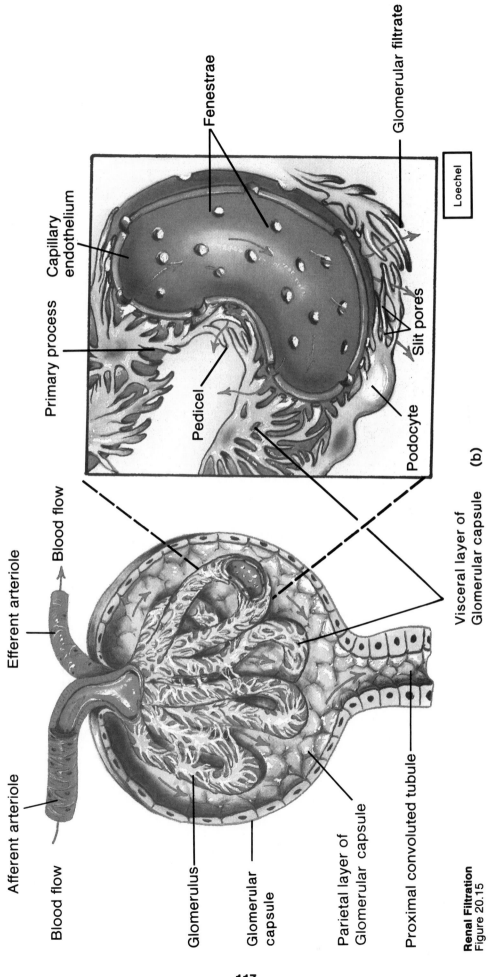

Fenestrae

Glomerular filtrate

Loechel

Capillary endothelium

Primary process

Pedicel

Slit pores

Podocyte

Visceral layer of Glomerular capsule

(b)

Blood flow

Blood flow

Efferent arteriole

Afferent arteriole

Glomerulus

Glomerular capsule

Parietal layer of Glomerular capsule

Proximal convoluted tubule

Renal Filtration
Figure 20.15

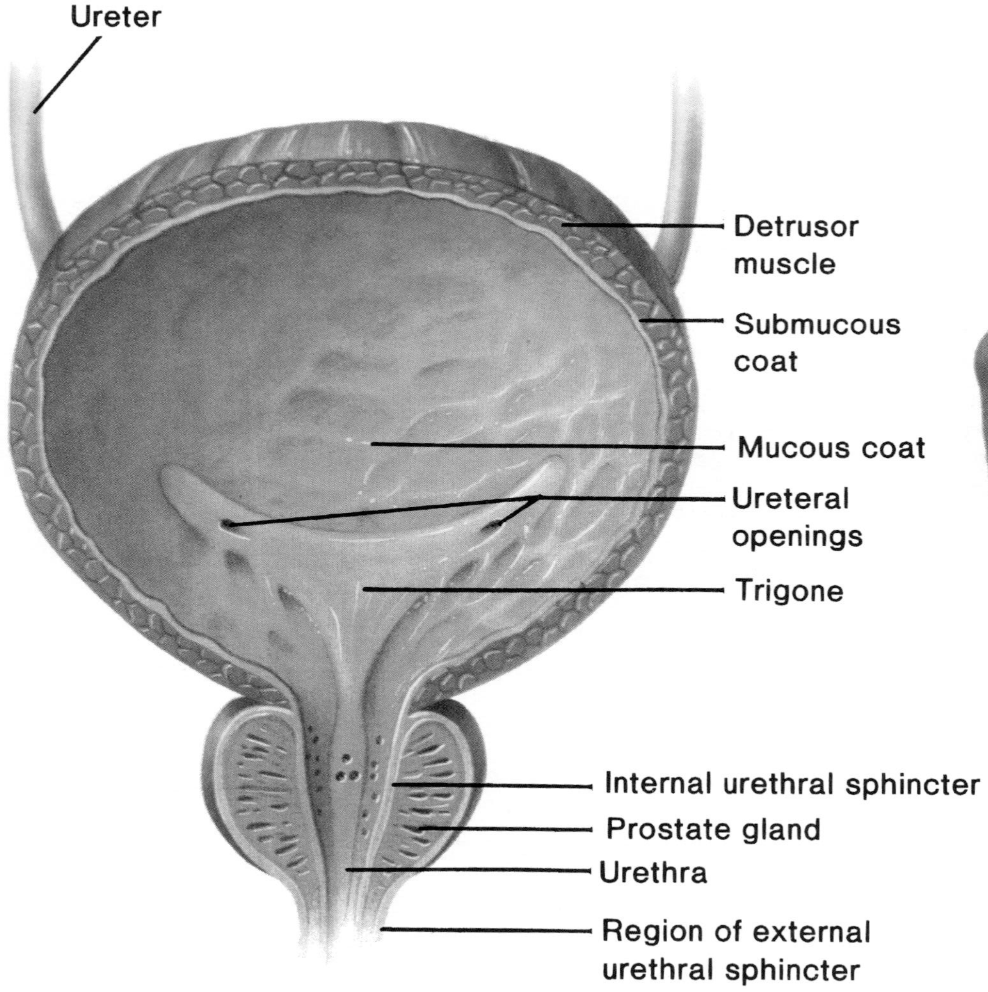

Ureter

Detrusor muscle

Submucous coat

Mucous coat

Ureteral openings

Trigone

Internal urethral sphincter

Prostate gland

Urethra

Region of external urethral sphincter

(a)

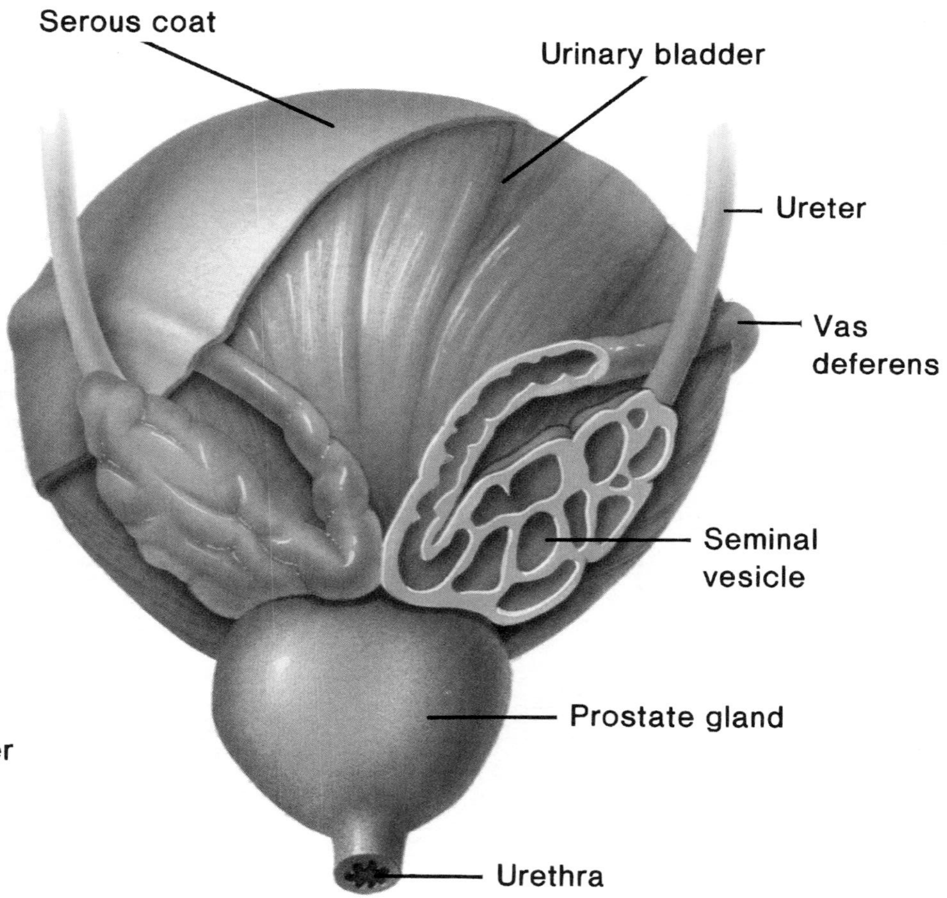

Ureter

Urinary bladder

Vas deferens

Seminal vesicle

Prostate gland

Urethra

Serous coat

(b)

Buck

Urinary Bladder
Figure 20.29

118

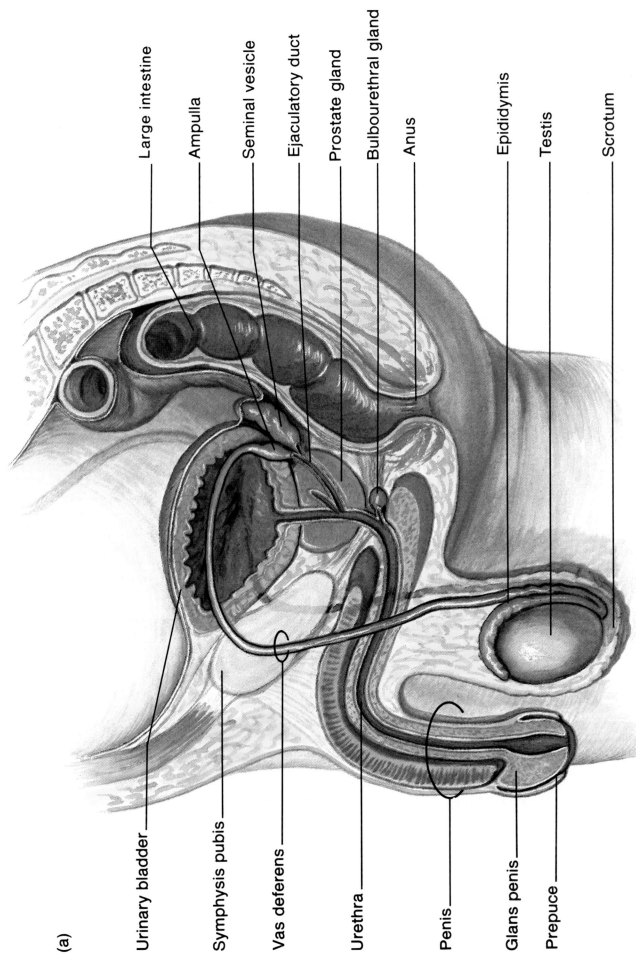

Large intestine

Ampulla

Seminal vesicle

Ejaculatory duct

Prostate gland

Bulbourethral gland

Anus

Epididymis

Testis

Scrotum

Urinary bladder

Symphysis pubis

Vas deferens

Urethra

Penis

Glans penis

Prepuce

(a)

Male Reproductive System
Figure 22.1a

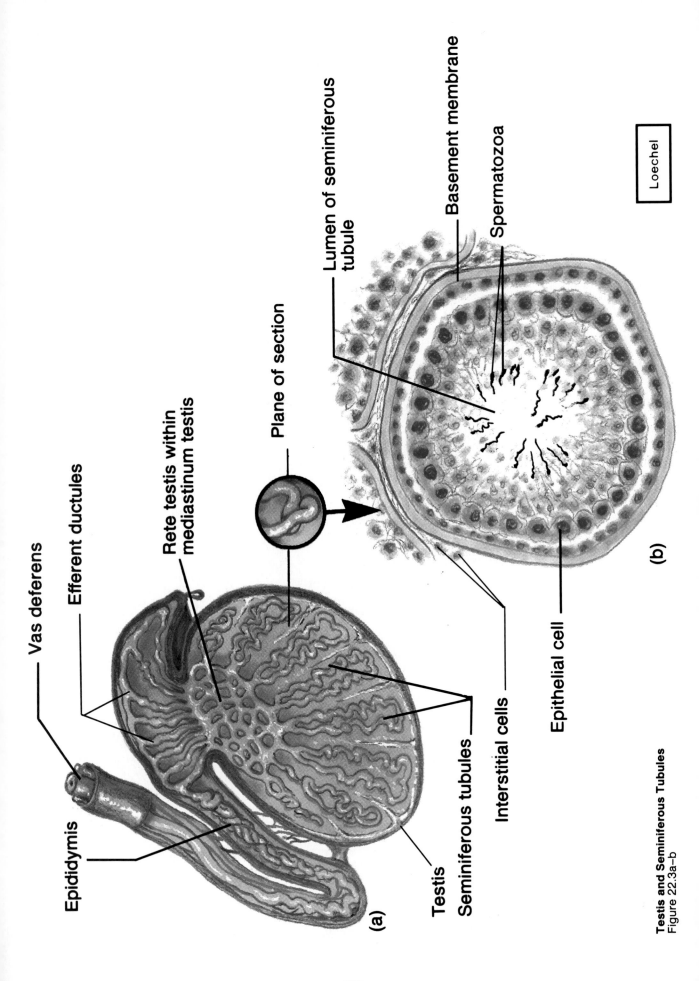

Vas deferens

Efferent ductules

Rete testis within
mediastinum testis

Epididymis

Plane of section

Lumen of seminiferous
tubule

Basement membrane

Spermatozoa

(a)

(b)

Loechel

Testis
Seminiferous tubules

Interstitial cells

Epithelial cell

Testis and Seminiferous Tubules
Figure 22.3a–b

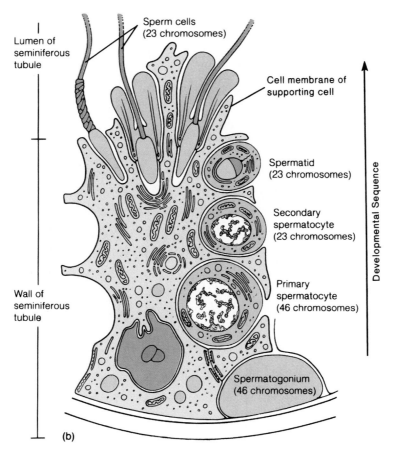

Lumen of
seminiferous
tubule

Sperm cells
(23 chromosomes)

Cell membrane of
supporting cell

Spermatid
(23 chromosomes)

Secondary
spermatocyte
(23 chromosomes)

Primary
spermatocyte
(46 chromosomes)

Wall of
seminiferous
tubule

Spermatogonium
(46 chromosomes)

Developmental Sequence

(b)

Spermatogenesis
Figure 22.5b

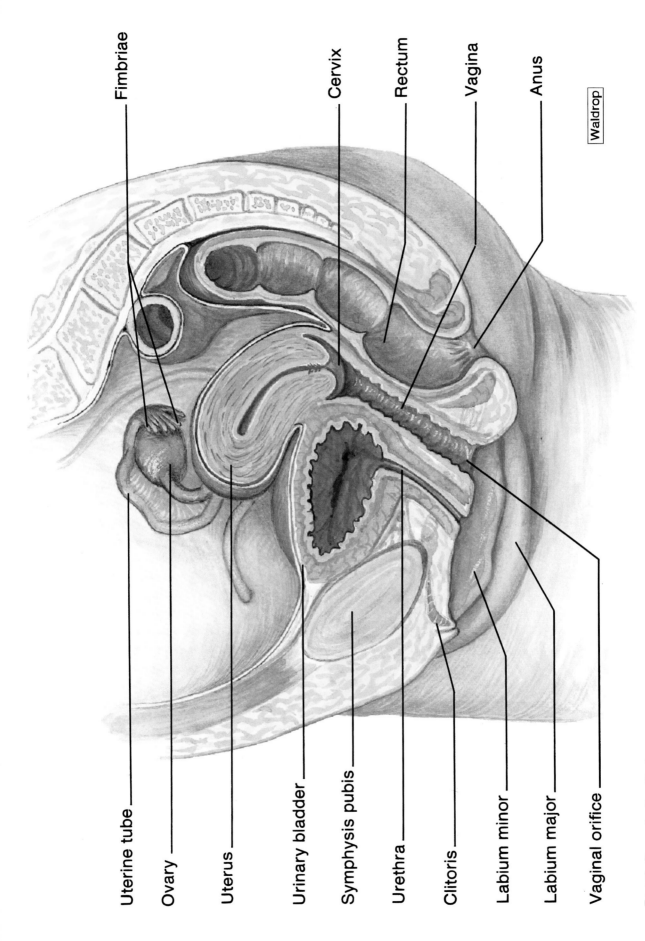

Fimbriae

Cervix

Rectum

Vagina

Anus

Waldrop

Uterine tube

Ovary

Uterus

Urinary bladder

Symphysis pubis

Urethra

Clitoris

Labium minor

Labium major

Vaginal orifice

Female Reproductive System
Figure 22.16a

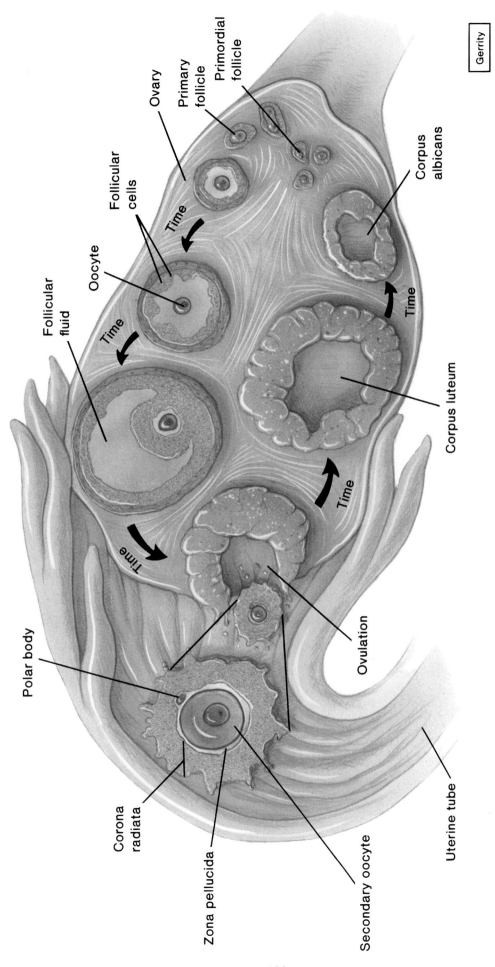

Ovary

Primary follicle

Primordial follicle

Follicular cells

Oocyte

Time

Follicular fluid

Time

Time

Polar body

Time

Corona radiata

Zona pellucida

Secondary oocyte

Ovulation

Uterine tube

Corpus albicans

Time

Corpus luteum

Gerrity

Ovarian Cycle
Figure 22.22

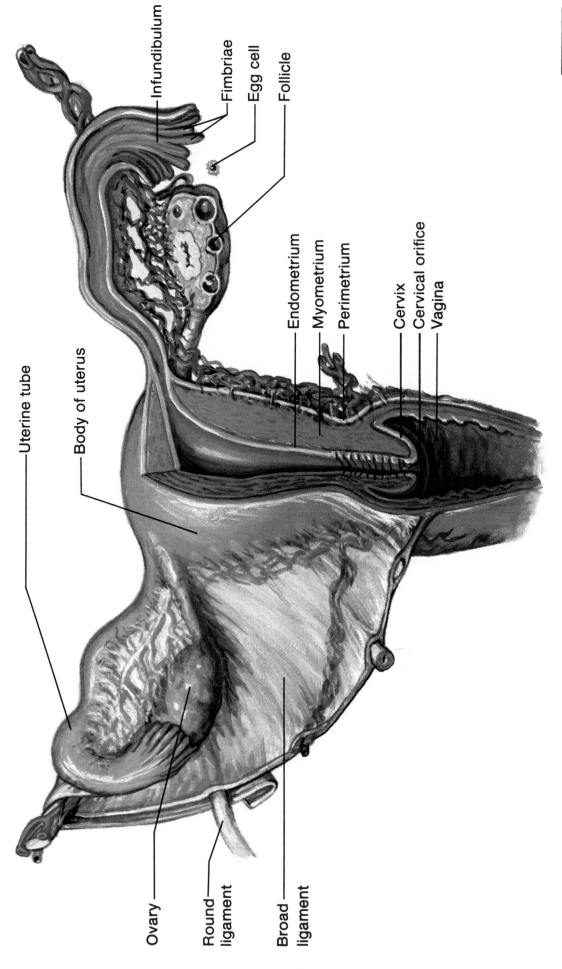

Infundibulum

Fimbriae

Egg cell

Follicle

Endometrium

Myometrium

Perimetrium

Cervix

Cervical orifice

Vagina

Uterine tube

Body of uterus

Ovary

Round
ligament

Broad
ligament

Female Internal Reproductive Organs
Figure 22.24

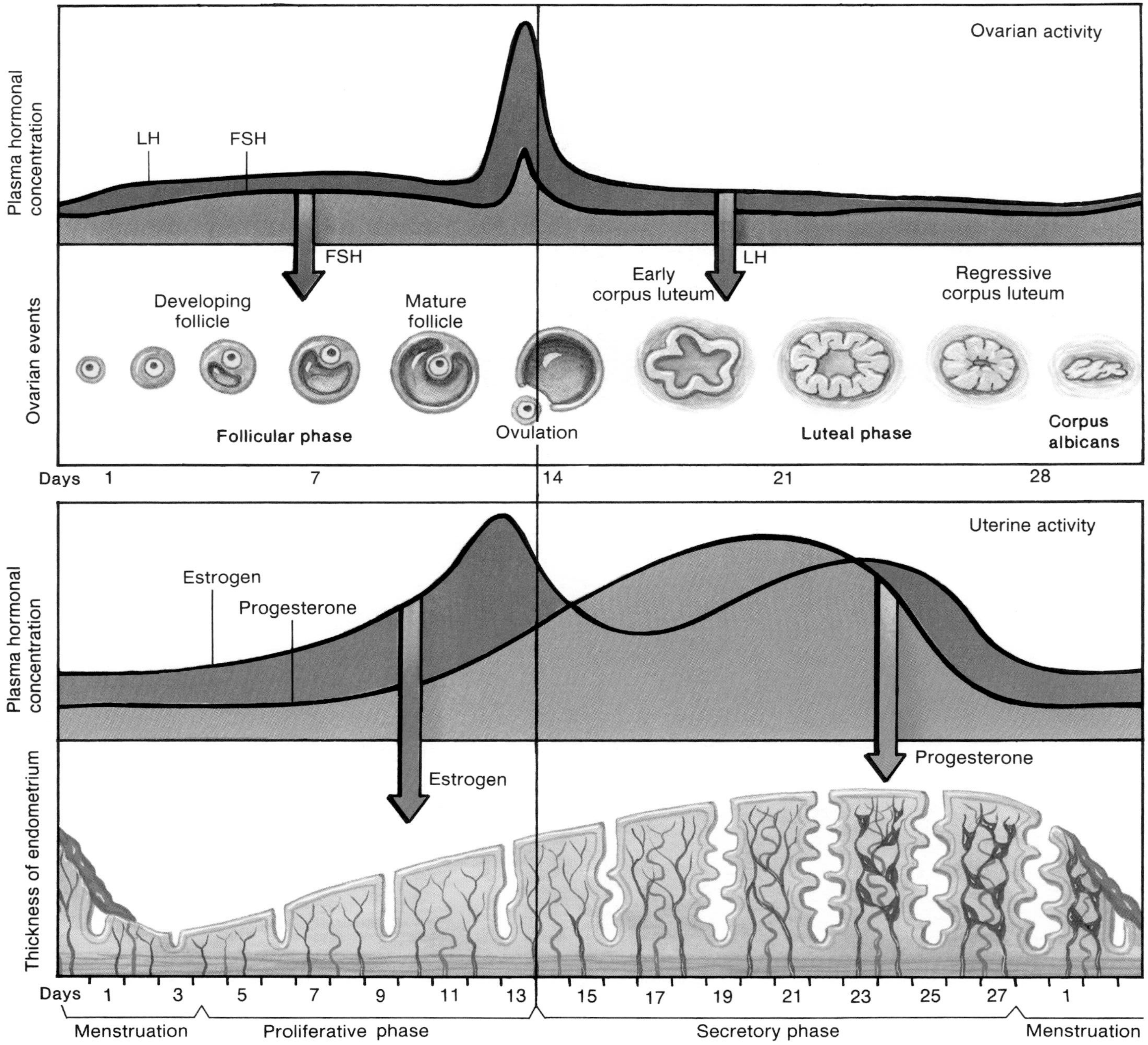

Menstrual Cycle
Figure 22.29

125

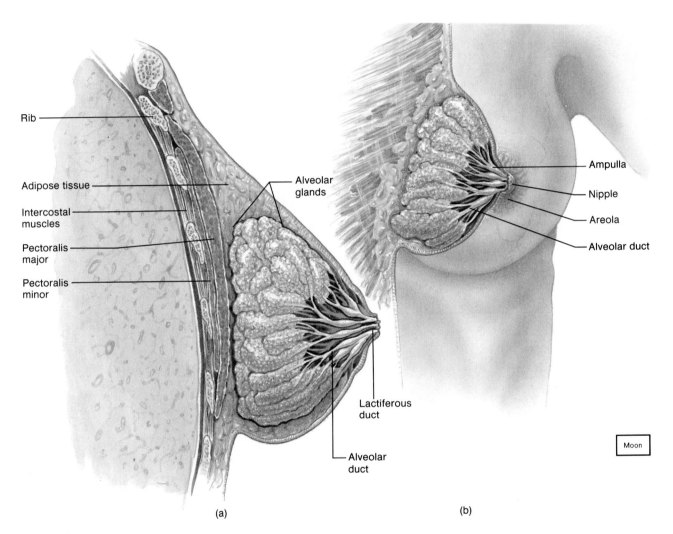

Rib

Adipose tissue

Intercostal
muscles

Pectoralis
major

Pectoralis
minor

Alveolar
glands

Lactiferous
duct

Alveolar
duct

Ampulla

Nipple

Areola

Alveolar duct

Moon

(a)

(b)

Breast Structure
Figure 22.38a–b

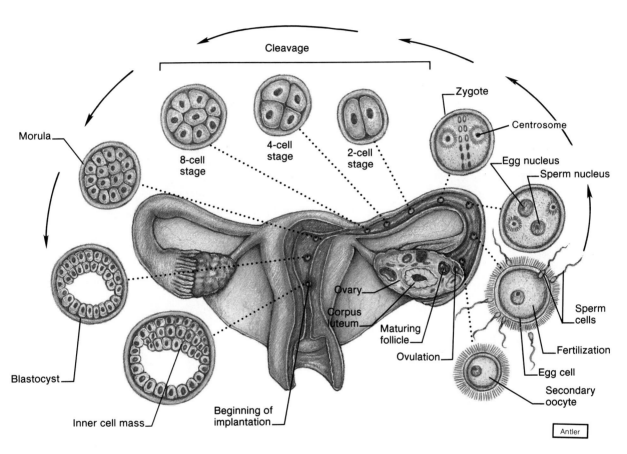

Human Embryonic Development
Figure 22.33

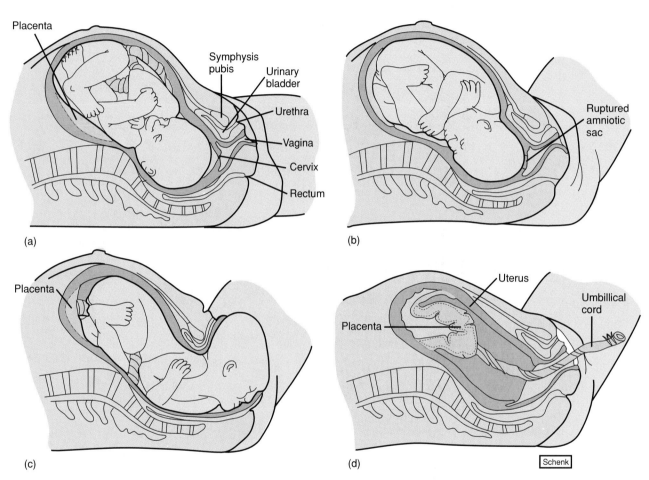

(a)

Placenta

Symphysis pubis

Urinary bladder

Urethra

Vagina

Cervix

Rectum

(b)

Ruptured amniotic sac

(c)

Placenta

(d)

Uterus

Placenta

Umbillical cord

Schenk

Birth Process
Figure 22.37

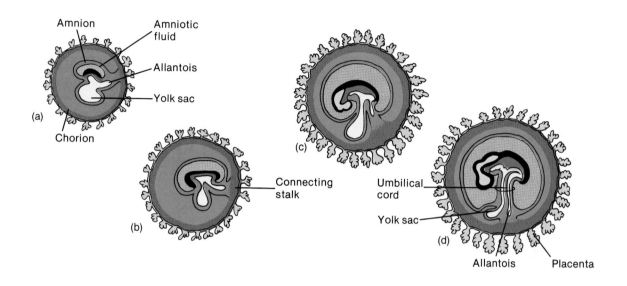

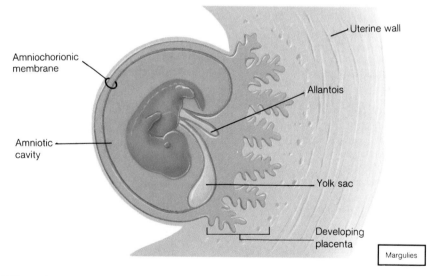

Formation of Placenta
Figures 23.20a–c and 23.21

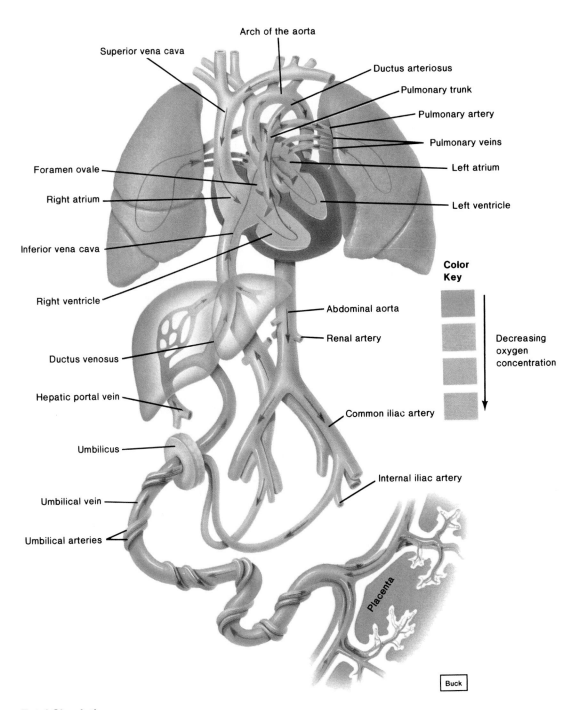

Arch of the aorta

Superior vena cava

Ductus arteriosus

Pulmonary trunk

Pulmonary artery

Pulmonary veins

Left atrium

Foramen ovale

Right atrium

Left ventricle

Inferior vena cava

Right ventricle

Abdominal aorta

Renal artery

Ductus venosus

Hepatic portal vein

Common iliac artery

Umbilicus

Internal iliac artery

Umbilical vein

Umbilical arteries

Placenta

Color
Key

Decreasing
oxygen
concentration

Buck

Fetal Circulation
Figure 23.28

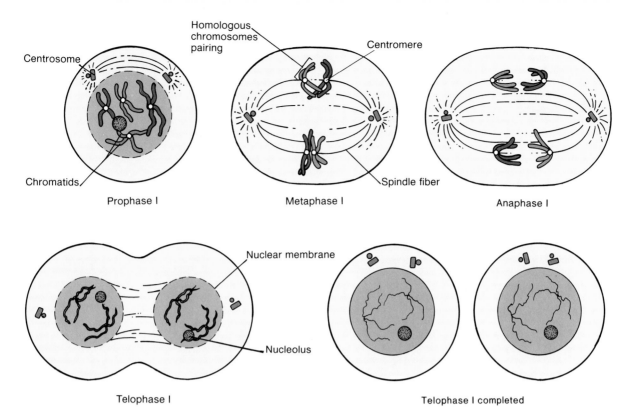

First Meiotic Division
Figure 24.9

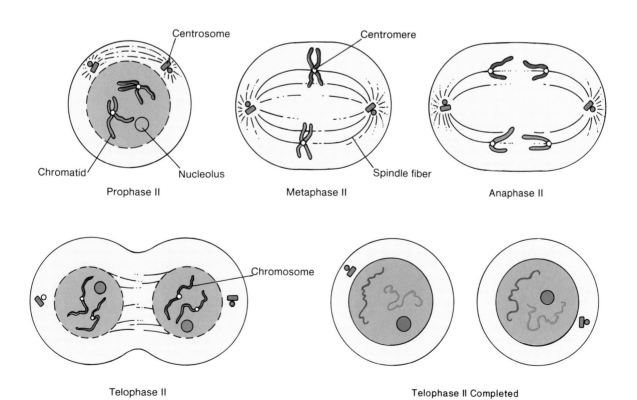

Second Meiotic Division
Figure 24.11

CREDITS

Line Art

Fig. 4.27: From Stuart Ira Fox, *Human Physiology,* 3d ed. Copyright © 1990 Wm. C. Brown Communications, Inc., Dubuque, Iowa. All Rights Reserved. Reprinted by permission.

Fig. 7.18a: From Kent M. Van De Graaff, *Human Anatomy,* 3d ed. Copyright © 1992 Wm. C. Brown Communications, Inc., Dubuque, Iowa. All Rights Reserved. Reprinted by permission.

Fig. 7.18b: From Kent M. Van De Graaff, *Human Anatomy,* 3d ed. Copyright © 1992 Wm. C. Brown Communications, Inc., Dubuque, Iowa. All Rights Reserved. Reprinted by permission.

Fig. 7.20: From Kent M. Van De Graaff, *Human Anatomy,* 3d ed. Copyright © 1992 Wm. C. Brown Communications, Inc., Dubuque, Iowa. All Rights Reserved. Reprinted by permission.

Fig. 7.23: From Kent M. Van De Graaff, *Human Anatomy,* 3d ed. Copyright © 1992 Wm. C. Brown Communications, Inc., Dubuque, Iowa. All Rights Reserved. Reprinted by permission.

Fig. 7.28: From Kent M. Van De Graaff, *Human Anatomy,* 3d ed. Copyright © 1992 Wm. C. Brown Communications, Inc., Dubuque, Iowa. All Rights Reserved. Reprinted by permission.

Fig. 9.20: From Kent M. Van De Graaff and Stuart Ira Fox, *Concepts of Human Anatomy and Physiology,* 3d ed. Copyright © 1992 Wm. C. Brown Communications, Inc., Dubuque, Iowa. All Rights Reserved. Reprinted by permission.

Fig. 9.21: From Kent M. Van De Graaff and Stuart Ira Fox, *Concepts of Human Anatomy and Physiology,* 3d ed. Copyright © 1992 Wm. C. Brown Communications, Inc., Dubuque, Iowa. All Rights Reserved. Reprinted by permission.

Fig. 11.12: From Kent M. Van De Graaff, *Human Anatomy,* 3d ed. Copyright © 1992 Wm. C. Brown Communications, Inc., Dubuque, Iowa. All Rights Reserved. Reprinted by permission.

Fig. 11.32: From Kent M. Van De Graaff, *Human Anatomy,* 3d ed. Copyright © 1992 Wm. C. Brown Communications, Inc., Dubuque, Iowa. All Rights Reserved. Reprinted by permission.

Fig. 11.36: From Kent M. Van De Graaff and Stuart Ira Fox, *Concepts of Human Anatomy and Physiology,* 2d ed. Copyright © 1989 Wm. C. Brown Communications, Inc., Dubuque, Iowa. All Rights Reserved. Reprinted by permission.

Fig. 11.38: From Kent M. Van De Graaff, *Human Anatomy,* 3d ed. Copyright © 1992 Wm. C. Brown Communications, Inc., Dubuque, Iowa. All Rights Reserved. Reprinted by permission.

Fig. 12.15b: From Stuart Ira Fox, *Human Physiology,* 4th ed. Copyright © 1993 Wm. C. Brown Communications, Inc., Dubuque, Iowa. All Rights Reserved. Reprinted by permission.

Fig. 13.13: From Kent M. Van De Graaff and Stuart Ira Fox, *Concepts of Human Anatomy and Physiology,* 3d ed. Copyright © 1992 Wm. C. Brown Communications, Inc., Dubuque, Iowa. All Rights Reserved. Reprinted by permission.

Fig. 14.1: From Kent M. Van De Graaff, *Human Anatomy,* 3d ed. Copyright © 1992 Wm. C. Brown Communications, Inc., Dubuque, Iowa. All Rights Reserved. Reprinted by permission.

Fig. 14.3: From Kent M. Van De Graaff and Stuart Ira Fox, *Concepts of Human Anatomy and Physiology,* 3d ed. Copyright © 1992 Wm. C. Brown Communications, Inc., Dubuque, Iowa. All Rights Reserved. Reprinted by permission.

Fig. 14.31: From Kent M. Van De Graaff, *Human Anatomy,* 3d ed. Copyright © 1992 Wm. C. Brown Communications, Inc., Dubuque, Iowa. All Rights Reserved. Reprinted by permission.

Fig. 14.48: From Kent M. Van De Graaff and Stuart Ira Fox, *Concepts of Human Anatomy and Physiology,* 3d ed. Copyright © 1992 Wm. C. Brown Communications, Inc., Dubuque, Iowa. All Rights Reserved. Reprinted by permission.

Fig. 16.12: From Kent M. Van De Graaff, *Human Anatomy,* 3d ed. Copyright © 1992 Wm. C. Brown Communications, Inc., Dubuque, Iowa. All Rights Reserved. Reprinted by permission.

Fig.18.34: From Kent M. Van De Graaff and Stuart Ira Fox, *Concepts of Human Anatomy and Physiology,* 3d ed. Copyright © 1992 Wm. C. Brown Communications, Inc., Dubuque, Iowa. All Rights Reserved. Reprinted by permission.

Fig. 20.11: From Kent M. Van De Graaff and Stuart Ira Fox, *Concepts of Human Anatomy and Physiology,* 3d ed. Copyright © 1992 Wm. C. Brown Communications, Inc., Dubuque, Iowa. All Rights Reserved. Reprinted by permission.

Fig. 22.3: From Kent M. Van De Graaff, *Human Anatomy,* 2d ed. Copyright © 1988 Wm. C. Brown Communications, Inc., Dubuque, Iowa. All Rights Reserved. Reprinted by permission.

Fig. 22.5b: From Kent M. Van De Graaff and Stuart Ira Fox, *Concepts of Human Anatomy and Physiology,* 3d ed. Copyright © 1992 Wm. C. Brown Communications, Inc., Dubuque, Iowa. All Rights Reserved. Reprinted by permission.

Fig. 22.24: From Kent M. Van De Graaff and Stuart Ira Fox, *Concepts of Human Anatomy and Physiology,* 3d ed. Copyright © 1992 Wm. C. Brown Communications, Inc., Dubuque, Iowa. All Rights Reserved. Reprinted by permission.

Fig. 22.37: From Kent M. Van De Graaff and Stuart Ira Fox, *Concepts of Human Anatomy and Physiology,* 3d ed. Copyright © 1992 Wm. C. Brown Communications, Inc., Dubuque, Iowa. All Rights Reserved. Reprinted by permission.

Fig. 22.38a–b: From Kent M. Van De Graaff and Stuart Ira Fox, *Concepts of Human Anatomy and Physiology,* 3d ed. Copyright © 1992 Wm. C. Brown Communications, Inc., Dubuque, Iowa. All Rights Reserved. Reprinted by permission.

Photographs

Fig. 5.1b: © Edwin Reschke

Fig. 5.2b: © Edwin Reschke

Fig. 5.3b: © Manfred Kage/Peter Arnold, Inc.

Fig. 5.5b: © Edwin Reschke

Fig. 5.6b: © Fred Hossler/Visuals Unlimited

Fig. 5.15b: © Edwin Reschke

Fig. 5.16b: © Edwin Reschke

Fig. 5.17b: © Edwin Reschke

Fig. 5.20b: © Edwin Reschke

Fig. 5.21b: © Edwin Reschke

Fig. 5.22b: © John Cunningham/Visuals Unlimited

Fig. 5.23b: © Victor B. Eichler

Fig. 5.24b: © Edwin Reschke

Fig. 5.25b: © Edwin Reschke

Fig. 5.26b: © Edwin Reschke

Fig. 5.27b: © Manfred Kage/Peter Arnold, Inc.

Fig. 5.28b: © Manfred Kage/Peter Arnold, Inc.